Trends in Software Testing

Hrushikesha Mohanty · J.R. Mohanty
Arunkumar Balakrishnan
Editors

Trends in Software Testing

Springer

Editors
Hrushikesha Mohanty
School of Computer and Information Sciences
University of Hyderabad
Hyderabad, Telengana
India

J.R. Mohanty
School of Computer Applications
KIIT University
Bhubaneswar, Odisha
India

Arunkumar Balakrishnan
Mettler Toledo Turing Softwares
Coimbatore, Tamil Nadu
India

ISBN 978-981-10-9351-7 ISBN 978-981-10-1415-4 (eBook)
DOI 10.1007/978-981-10-1415-4

Softcover reprint of the hardcover 1st edition 2016

Printed on acid-free paper

This Springer imprint is published by Springer Nature
The registered company is Springer Science+Business Media Singapore Pte Ltd.

Preface

Software testing is an important phase in the software-development cycle. Well-tested software that is free of bugs is the goal of the software developer and the expectation of the user. Today's world has "gone digital" and offers services to people for their ease and comfort. Many essential services are also being delivered online over the Internet. Thus, our life style is becoming increasingly dependent on gadgets and even more on the software that controls the gadgets. The malfunctioning of services offered online by way of these gadgets is what users least expect. For the purpose, well-tested services are the first requirement before software products appear on the market for consumer use.

Software testing broadly aims to certify not only the accuracy of the logic embedded in code but also adherence to functional requirements. Traditional testing strives for verification of these two aspects. Academia as well as industry have been working on software testing in their own ways and have contributed to the body of research on testing algorithms and practices. As a result, tools exist to automate software-testing efforts so that testing can be accomplished in a faster and less-expensive manner. However, as new computing paradigms and platforms emerge, so does the need to re-examine software testing. Now both academia and industry are searching for solutions to the problems currently faced by software development. Some of these current challenges are addressed in this book.

The challenges software testing faces now require an integrative solution; this means that new solutions must run alongside all traditional solutions. However, well-developed and well-practiced concepts should not be abandoned in search of the new. New issues are arising as changes in technology and varied applications are identified and dealt with zeal by both academia and industry professionals. For that reason, this book includes the views and findings of both academia and industry. The challenges they address include test debt, agile testing, security testing, uncertainty in testing, separation, software-system evolution, and testing as a service. In order to motivate readers, we will brief them on these issues and invite interested readers to explore chapters of their interest.

Software developers usually race against time to deliver software. This mainly happens in response to users who require fast automation of tasks they have been managing manually. The sooner these tasks are automated, the faster software developers reap financial gains. This has initiated unbelievable growth in the software industry. Software business houses compete with each other to reach out to users with the promise of deliverables as quickly as possible. In the process, developers take short cuts to meet deadlines. Short cuts can be made in each phase of the software-development cycle. Among all of the phases, the testing phase is most prone to short cuts being taken. Because quality software requires exhaustive testing, the process is time consuming. Although taking short cuts ensures that users' delivery expectations are met, they open the door for problems to creep in that appear at later time. These problems later may make things messy and even expensive for a company to fix. This phenomenon is termed "test debt." Test debt occurs due to inadequate test coverage and improper test design. The chapter, *Understanding Test Debt*, by SG Ganesh, Mahesh Muralidharan, and Raghu Kalyan Anna, takes up this issue and elaborates on defining technical debt as a whole and test debt specifically. As the authors say, the causes for technical debt could be many including work-schedule pressure, lack of well-trained personnel, lack of standards and procedures, and many other similar issues that bring in adhocism into software development. Test debt can occur during the requirement-engineering, design, coding, and testing phases. This chapter considers test-debt management from the authors' experience. The concepts presented here include quantifying test debt, paying off debt periodically, and preventing test debt. The best method to avoid test debt is to take appropriate measures, such as adopting best practices in coding and testing, so that test debt is prevented. Finally, the authors illustrate their ideas by quoting two case studies and ending with some suggestions such as the development of tools that can detect smells when debt occurs in software development, particularly during the testing phase.

Early detection of test smell is better because it restricts test debt from increasing. Thus, the idea of testing should start at an early stage of development; however, testing traditionally occurs at the end of coding. The recent idea is to develop in small steps and then test; the cycle is repeated to complete development of the software. Thus, test smell detection happens early in development. This process is known as "agile testing." One chapter, *Agile Testing* by Janakirama Raju Penmetsa, discusses methodology along with presenting case studies to familiarize beginners with this new concept. The hallmark of agile testing is collaboration, transparency, flexibility, and retrospection. In traditional testing methodology, the developer and the tester "chase" each other with negative attitudes: Developers find "devils" in testers, and testers try to pin down developers. This attitude provokes an unwanted working atmosphere. Agile method works on the principle of co-operation and transparency and thus results in a congenial working atmosphere.

The chapter on agile testing presents a brief discussion on the agile process by quoting extreme programming and scrum. A reader new to this concept gets a basic idea of the agile process and then explores the use of this concept in the realm of testing. The author, with the help of an illustration, draws a picture of the agile-testing process, which is based on an iterative and collaborative approach in software development. An application is divided into smaller atomic units with respect to user requirements. Each unit is developed, tested, and analysed by all concerned including a designer, a code developer, and a tester. In case there is any problem, issues are sorted out then and there instead of carrying test debt forward. Advantages and disadvantages of agile testing are also identified. The success of agile testing greatly depends on a new class of testing tools that automate the process of testing on the basis of the agile-testing process.

Traditionally functional testing has been a prime concern in software testing because this ensures users that requirements will be met. Among nonfunctional issues, responsiveness testing takes the first priority because users usually want a system to perform as quickly as possible. Recently the security of software systems has become the main priority because it is a prime need, particularly when systems become complex as distributed and loosely coupled systems. The integration of third-party software has become prevalent even more so in the case of large systems. This provides a great degree of vulnerability to systems it could many open doors for hackers to sneak through and create havoc in terms of system-output accuracy and performance. This underscores the urgency of security testing to ensure that such doors are closed. Security testing minimizes the risk of security breach and ensures the confidentiality, integrity, and availability of customer transactions. Another chapter, *Security Testing* by Faisal Anwer, Mohd. Nazir, and Khurram Mustafa, takes up the issue of security testing. On discussing the challenges involved, the chapter proceeds to detail the significance of security testing. The authors explore the life cycle of software development and argue that security concerns must be addressed at every single phase because security is not a one-time activity. Security in requirement, design, coding, and testing is to be ensured in order to minimise system vulnerability to security threats. Static security testing involves an automatic check of system vulnerability without the execution of programs, whereas dynamic security testing involves testing of programs while the gadget is running. Static-testing techniques include code review, model checking, and symbolic checking. Techniques such as fuzz testing, concolic testing, and search-based testing techniques come under the category of dynamic testing. The authors present a case study to illustrate the ideas of security testing. They argue that security testing must be embedded in the software-development process. Software security should be checked in a two-layer approach including checking at each phase of the software-development life cycle and again in integrated manner throughout the development life cycle. A discussion of industry practices followed in security testing is contributed, and the authors go forward to explore industry requirements and future trends. The chapter concludes with a positive remark expecting a bright future in software-quality assurance, particularly in the context of security testing, by adopting phase-based security testing instead of demonstration-based conventional testing.

Testing itself is a process consisting of test selection, test classification, test execution, and quality estimation. Each step of testing is dependant on many factors, some of which are probalistic, e.g., the selection of a test case from a large set of candidate test cases. In that case, there could be a compromise in test quality. Thus, uncertainty in software testing is an important issue to be studied. At each phase of software development, it is required to estimate uncertainty and take appropriate measures to prevent uncertainty. Because the greater the uncertainty, the farther away the actual product could be from the desired one. Because this issue is serious and not has been sufficiently addressed in software engineering, this book includes a chapter—*Uncertainty in Software Testing* by Salman Abdul Moiz—on the subject.

The chapter starts with a formal specification of a test case and introduces some sources of uncertainty. It discusses at length about what contributes to uncertainty. Uncertainty starts from uncertainty in requirement engineering and priortising user requirements. Uncertainty, if not eliminated early on, affects other activities, such as priortising test cases and their selections, at a later stage of software development. In the case of systems with inherent asynchrony, the results differ from one test run to another. Uncertainty plays a significant role in assuring the quality testing of such systems. Currently many advanced systems adopt heuristics for problem solving, and some use different machine-learning techniques. These also contribute to uncertainty regarding software output because these techniques depend on variables that could be stochastic in nature.

In order to manage uncertainty, there must be a way to measure it. The chapter reviews some of the important techniques used to model uncertainty. Based on the type of system, an appropriate model to measure uncertainty should be used. The models include Bayesian approach, fuzzy logic, HMM, and Rough sets. Machine-learning techniques are also proposed to determine uncertainty from previous test runs of similar systems. The chapter also provides rules of thumb for making design decisions in the presence of uncertainty.

Measuring uncertainty, ensuring security, adopting agile testing, and many such software-engineering tasks mostly follow a bottom-up approach. That is the reason why modularisation in software engineering is an important task. Modularisation aims at isolation with minimal dependence among modules. Nevertheless, the sharing of memory with the use of pointers creates inconsistency when one module changes value to a location without the knowledge of others who share the location. This book has a chapter, *Separation Logic to Meliorate Software Testing and Validation* by Abhishek Kr. Singh and Raja Natarajan, on this issue.

This chapter explores the difficulties in sharing of memory locations through presenting some examples. It introduces the notion of separation logic. This logic is an extension to Hoare logic. The new logic uses a programming language that has four commands for pointer manipulation. The commands perform the usual heap operations such as lookup, update, allocation, and deallocation. Formal syntax for each command is introduced. Sets of assertions and axioms are defined to facilitate reasoning for changes due to each operation. For this purpose, basic structural rules are defined; then for specific cases, a set of derived rules are also defined. This helps in fast reasoning because these rules are used directly instead of deriving the same from basic rules at the time of reasoning. Before and after each operation, the respective rules are annotated in the program to verify anomalies, if any, created by the operation. The idea is illustrated by a case study of list traversal. The chapter, while introducing separation logic, also proposes a formal language that naturally defines separation in memory usages. The technique proposed here presents a basic method to assure separation among modules while using common pointers. This helps in building systems by modules and reasoning over its correctness.

Top-down system building has been a practice well-accepted by software developers. Agile methodology inspires developers to build a system in modules by taking user concerns as a priority. The system building is iterative, and the system architecture evolves through iterations. At each stage, the system architecture must be rationalized to build confidence, which is required for the subsequent design, testing, and maintenance of complex systems. This is also required for the re-engineering of a system because re-engineering may result in changes to the system architecture. Considering the importance of stability study in evolving systems, this book includes a chapter on the topic, titled *mDSM: A Transformative Approach to Enterprise Software Systems Evolution*, by David Threm, Liguo Yu, SD Sudarsan, and Srini Ramaswamy. mDSM is an extension of the DSM (Design Structure Matrix) approach to software system design and testing. It was developed by the authors to address the design-driven rationalization of such complex software system architectures.

When modelling a complex system, three structural aspects are considered: instantiation, decomposition, and recombination. The first aspect aims to ensure that all design entities belong to the same- type domain. The second aspect considers that a system design can be refined into smaller subdomains. The third aspect assures a system can be reassembled from the constituting components. The authors propose that mDSM-enabling analysts view the software system in its entirety without losing sight of how the modules, units, subsystems, and components interact. Evolutionary-stability metrics is proposed to evaluate a software system's evolution at the lowest level and ensure that the software system can still be rationalized in its totality.

In the case of agile testing, the software is tested at each stage, and changes are made incrementally. At the end of each iteration, stability is computed to rationalize the evolved architecture. For this purpose, the authors introduce evolutionary-stability metrics for software systems and mDSM methodology. Essentially, the concept of normalized-compression distance (NCD) is introduced to study the difference between two versions of a software artifact. Metrics such as version stability, intercomponent version stability, branch stability, structure stability, and aggregate stability are the few ideas introduced in the chapter.

mDSM prescribes five steps to develop component-based architecture: (1) decomposition of a system into components; (2) documentation and understanding of interaction among the components; (3) calculating the evolutionary stability of the components; (4) laying out a mDSM matrix with the components labeled in rows and stability values presented in columns with the corresponding matrix cells; and (5) performing potential reintegration analysis. The concept of mDSM and its impact, particularly in testing, are illustrated by a number of case studies. The chapter ends with listing the advantages of mDSM as well as a concluding remarks on the future direction of research.

Software testing is becoming sophisticated as system complexity increases. This not only requires a large pool of experts but also requires sophisticated technology to perform qualitative testing. Further rapid changes in technology call for the augmentation of testing resources. At this point, many software houses find it difficult to maintain testing resources to meet ever-increasing challenges. At the same time, testing as a service is emerging. Corporate houses specializing on testing augment their resources to meet emerging testing requirements. This becomes a win–win business proposition among software developers as well as test-service providers. Again, with the advent of new technology, such as cloud computing, this service can be provided with optimal cost because test resources can be shared on the cloud by several service providers. Thus, testing as a service has not only become a feasible solution but also ready to create good returns on investments. Corporates are engaging in making testing happen on a large scale. Considering the impact of this fledgling concept, the book has included a chapter on the topic, *Testing as a Service*, by Pankhuri Mishra and Neeraj Tripathi.

The chapter lists difficulties in traditional testing: It is expensive, inflexible in resource sharing, and demands high-level testing experts. This makes a case for testing as a service by a service provider with expertise in software testing. The authors propose a generic model for such a service and define the roles of test-service providers and consumers. Service providers aim for quality testing, efficiency, and good returns. At the same time, they must exercise caution for testing services. Among the main concerns, security is crucial because service consumers are required to provide input patterns to testers, whereas from a business-interest point of view, these input patterns may contain many business secrets. High-level abstraction of testing-service architecture is proposed. A workflow depicting the chain of actions required to create a service is proposed. A higher-level implementation of this architecture familiarizes with a tool that automates services. The working of the architecture is explained with an example. On making a comment on the pricing of testing service, the authors hint at the usability of cloud technology for providing test services. They base this idea on the suitability of the technology because it helps in the optimal sharing of resources. The chapter ends with a concluding remark emphasizing standardization of this new upcoming service.

This book intends to give a broad picture of trends in software testing. For this purpose, we have included seven chapters addressing different issues that are currently being addressed by testing professionals. We do not claim totality in presenting the trends in software testing, but the chosen topics cover some important issues. Of seven chapters, four are contributed by practitioners from different software houses engaged in testing. The other three chapters from academia add rigour to the exploration of the issues. We hope that this book will be useful to students, researchers, and practitioners with an interest in software testing.

Hrushikesha Mohanty
J.R. Mohanty
Arunkumar Balakrishnan

Acknowledgments

The genesis of this book goes to 12th International Conference on Distributed Computing and Internet Technology (ICDCIT) held in February 2016. Software testing was a theme for the industry symposium held as a prelude to the main conference. In preparation of this book we received help from different quarters. I (Hrushikesha Mohanty) express my sincere thanks to the School of Computer and Information Sciences, University of Hyderabad for providing excellent environment for carrying out this work. I also extend my sincere thanks to Dr. Achyuta Samanta, Founder KIIT University for his inspiration and graceful support for hosting the ICDCIT series of conferences. Shri. D.N. Dwivedy of KIIT University deserves special thanks for making it happen. The help from ICDCIT organization committee members of KIIT University are thankfully acknowledged. Jnyanranjan Mohanty and Arunkumar extend their thanks to their respective organizations KIIT University and Mettler Toledo Turing Software.

Our special thanks to chapter authors who, despite their busy schedules, could contribute chapters for this book. We are also thankful to Springer for publishing this book. Particularly, for their support and consideration for the issues we have been facing while preparing the manuscript.

Contents

About the Editors

Hrushikesha Mohanty is currently a professor at School of Computer and Information Sciences, University of Hyderabad, Hyderabad, India. He has received his Ph.D. from IIT Kharagpur. His research interests include Distributed Computing, Software Engineering and Computational Social Science. Before joining University of Hyderabad, he worked at Electronics Corporation of India Limited for developing strategic real-time systems. He has published around 96 research papers and edited 6 books on computer science.

J.R. Mohanty is an associate professor in School of Computer Application, KIIT University, Bhubaneswar, Odisha, India. He has been working in the field of computer applications for 19 years. He earned his Ph.D. in Computer Science from Utkal University, India. His research interests include queuing networks, computational intelligence and data mining.

Arunkumar Balakrishnan is a senior consultant at M/s Mettler Toledo Turing Softwares, Coimbatore, India, for the past 12 years. He is also Professor & Director MCA in the Department of Computer Technology & Applications at Coimbatore Institute of Technology, Coimbatore, India, for the past 18 years (on half sabbatical since June 2003). He acquired his Ph.D. in Computer Science from Bharathiar University in 1996. He has published several papers in international journals and conference proceedings. The publications have been in the areas of Machine learning, Student modelling and Automated Software Testing.

About the Book

This book aims to highlight the current areas of concern in software testing. The contributors hail from academia as well as the industry to ensure a balanced perspective. Authors from academia are actively engaged in research in software testing, whereas those from the industry have first-hand experience in coding and testing. This book attempts to identify trends and some recurring issues in software testing, especially as pertaining to test debt, agile testing, security testing and uncertainty in testing, separation modules, design structure matrix and testing as a service.

Taking short cuts in system testing results in test debt and the process of system development must lead to reduce such debts. Keeping this in view, the agile testing method is explored. Considering user concerns on security and the impact that uncertainty makes in software system development, two chapters are included on these topics. System architecture plays an important role in system testing. Architectural components are required to be non-interfering and keep together for making a stable system. Keeping this in view chapters on separation and design structure matrix are included in the book. Finally, a chapter on testing as a service is also included. In addition to introducing the concepts involved, the authors have made attempts to provide leads to practical realization of these concepts. With this aim, they have presented frameworks and illustrations that provide enough hints towards system realization.

This book promises to provide insights to readers having varied interest in software testing. It covers an appreciable spread of the issues related to software testing. And every chapter intends to motivate readers on the specialties and the challenges that lie within. Of course, this is not a claim that each chapter deals with an issue exhaustively. But we sincerely hope that both conversant and novice readers will find this book equally interesting.

We hope this book is useful for students, researchers and professionals looking forward to explore the frontiers of software testing.

Understanding Test Debt

Ganesh Samarthyam, Mahesh Muralidharan and Raghu Kalyan Anna

Abstract Technical debt occurs when teams knowingly or unknowingly make technical decisions in return for short-term gain(s) in their projects. The test dimension of technical debt is known as test technical debt (or test debt). Test debt is an emerging topic and has received considerable interest from software industry in the last few years. This chapter provides an overview of test debt, factors that contribute to test debt, and strategies for repaying test debt. The chapter also discusses how to identify "test smells" and refactor them for repaying technical debt in industrial projects using numerous examples and case studies. This chapter would be of considerable value to managers and leads working in IT companies as well as researchers working in the area of test debt.

Keywords Software testing · Technical debt · Test debt · Test smells · Test code · Refactoring test cases · Test design · Effective testing · Testing approaches · Test debt case studies

1 Introduction

Technical debt occurs when teams knowingly or unknowingly make technical decisions in return for short-term gains in their projects. The test dimension of technical debt is known as "test technical debt", or "test debt" for short. Technical

G. Samarthyam (✉)
Bangalore, India
e-mail: sgganesh@gmail.com

M. Muralidharan
Siemens Research and Technology Center, Bangalore, India
e-mail: M.Mahesh@siemens.com

R.K. Anna
Symantec Software, Pune, India
e-mail: raghukalyan@gmail.com

H. Mohanty et al. (eds.), *Trends in Software Testing*,
DOI 10.1007/978-981-10-1415-4_1

debt—along with its various dimensions—has become increasingly relevant for organizations that develop and maintain large software systems.

This chapter provides an overview of test debt and is organized as follows: After providing a brief introduction to technical debt, this chapter presents a detailed discussion on factors that contribute to test debt, key strategies for managing test debt, two case studies on managing test debt in real-world projects, and concludes by identifying future directions on test debt for software testing community.

1.1 Technical Debt

Technical debt is the debt that accrues when we knowingly or unknowingly make wrong or non-optimal technical decisions [1]. Ward Cunningham coined the term "technical debt" [2] wherein long term software quality is traded for short-term gains in development speed. He used the metaphor of financial debt to indicate how incurring debt in the short run is beneficial but hurts badly in the long run if not repaid.

We incur financial debt when we take a loan. The debt does not cause problems as long as we repay the debt. However, if we do not repay debt, the interest on the debt keeps building over a period of time and finally we may land in a situation where we are not in a position to repay the accumulated financial debt leading to financial bankruptcy. Technical debt is akin to financial debt. When we take shortcuts during the development process either knowingly or unknowingly, we incur debts. Such debts do not cause problems when we repay the debt by performing refactoring to address the shortcut. If we keep accumulating technical debts without taking steps to repay it, the situation leads the software to "technical bankruptcy".

A high technical debt in a software project impacts the project in multiple ways. Ward Cunningham talks about the impact of technical debt on engineering organizations in this original paper [2]: "Entire engineering organizations can be brought to a stand-still under the debt load …". Israel Gat warns about technical debt [3]: "If neglected, technical debt can lead to a broad spectrum of difficulties, from collapsed roadmaps to an inability to respond to customer problems in a timely manner." Specifically, the Cost of Change (CoC) increases with increasing technical debt [4] that leads to poor productivity. In addition, it starts the vicious cycle where poor productivity leads to more focus on features (rather than associated quality) in the limited time, and that leads to more technical debts due to time constraints. Apart from technical consequences, high technical debt impacts the morale and motivation of the development teams in a negative manner [1].

There are several well-known causes that lead to technical debt, including:

- Schedule pressures.
- Lack of skilled engineers.
- Lack of documentation.

- Lack of process and standards.
- Lack of test suite.
- Delayed refactoring.
- Lack of knowledge.

Li et al. [5] list technical debt along the dimensions of requirements, architecture, design, code, test, build, documentation, infrastructure, and versioning. Figure 1 lists prominent dimensions of technical debt with examples. Traditionally, technical debt arising due to code debt received considerable attention [6]. Recently, design debt [1] as well as architecture debt [6, 7] have started to receive interest both from academia and industry. For overall pragmatic management of technical debt and project success, it is important to manage other dimensions of debt such as "infrastructure debt" and "test debt".

Although, there have been a few attempts to understand test debt [8, 9], the dimension has not been explored well. Given the vital role that testing plays in the software development process and the impact test debt can have on software projects, test debt and corresponding management techniques deserve more focus for developing high-quality software in industrial contexts.

Technical test debt (or test debt for short) occurs due to shortcuts (i.e., wrong or non-optimal decisions) related to testing. It is the test dimension of technical debt. For example, when unit tests are entirely missing (this scenario is common in legacy projects), it contributes to test debt. It is because not performing unit testing is a form of shortcut: though there is a short-term benefit of speeding-up development, there is a long term impact associated with that decision.

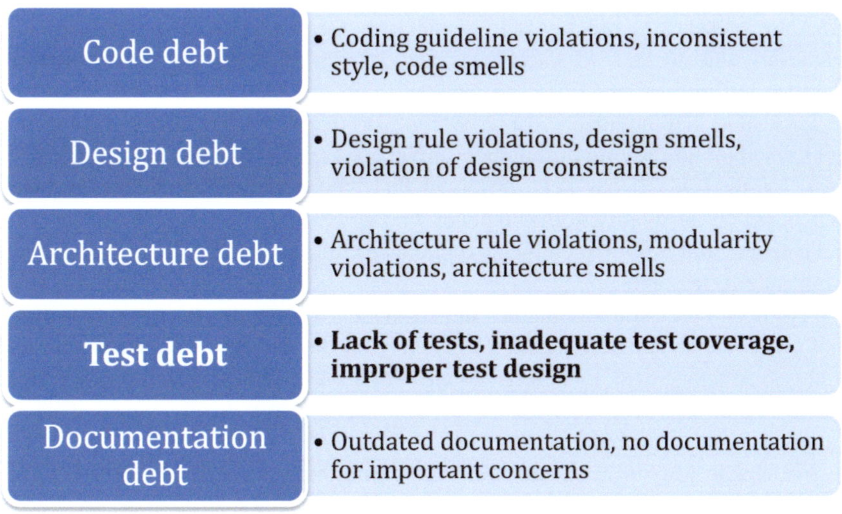

Fig. 1 Some of the prominent dimensions of technical debt with examples

1.2 Importance of Test Debt

Let us consider a scenario frequently observed in startup organizations that leads to test debt and subsequently makes the organization struggle for survival due to high test debt. Consider that a startup aims to release their product first in the market to get the "first mover advantage". To make that possible, the organization focuses exclusively on developing features. But, when it comes to testing, developers quickly test the most common scenarios on their machines rather than carrying out extensive and comprehensive testing. With heroic efforts, the team might be able to successfully develop and deliver a working product in a relatively short span of time and even end-up garnering considerable market share. However, the initial success is often not sustainable. While developing the first version, the startup has incurred huge amount of technical debt. Specifically in the test dimension, the team might have not created and performed automated tests or the carried out testing might have resulted in very poor test coverage. For the next version, the startup must focus on repaying the test debt by investing heavily on testing. Any delays or insufficient focus on repaying the debt would make the product unreliable. If the incurred debt is enormous, the development will come to a stand-still and no more progress could be made reliably, i.e., result in "technical bankruptcy" and may force the startup to abandon the product.

1.3 General Causes of Test Debt

The chapter has discussed already a few typical causes of technical debt and same contribute to test debt as well. Software teams are typically under tremendous schedule pressure to quickly deliver value to their stakeholders. Often, in such situations, due to lack of knowledge and awareness about test debt, the teams sacrifice testing best practices and incur test debt.

Lack of skilled engineers also contributes to test debt. Developers are typically not trained on unit testing during their university degree and often learn writing unit tests "on-the-job". It has been observed that "the quality of unit tests is mainly dependent on the quality of the engineer who wrote the tests" [10]. The observation indicate that unit tests written by inexperienced engineers are harder to maintain and contribute to test debt.

"Number obsession" is another not-so-evident but common cause of test debt that occurs when software teams focus solely on the numbers associated with tests such as coverage and total tests count. Their obsession is just to satisfy their management; however, in reality they sacrifice the test quality while achieving "numbers" and contribute to test debt.

Inadequate test infrastructure tools also contributes to test debt. For instance, consider that a test team uses an out-of-date test automation framework (i.e., framework that is nearing end-of-life or obsolete). The more the team postpones replacing the framework with a new framework, more tests are written in the older set-up. So, it costs more to upgrade to the newer versions and hence contribute to accumulating test debt.

Additionally, there are many other causes of test debt, such as inadequate test planning and estimation errors in test effort planning.

2 Classification of Test Debt

There are numerous aspects that contribute to test debt. We propose a classification of test debt based on the scope and type of testing performed[1] (Fig. 2).

2.1 Unit Testing

There are numerous short-cuts in unit testing that contribute to test debt (many of them covered by Deursen et al. [11]). Here we cover a few important aspects relating to unit testing that contribute to test debt.

- ***Inadequate unit tests***. Consider the scenario in which unit tests are inadequate or missing in a project. In such cases, the defects remain latent for longer time and may get caught at higher level tests (such as system tests). This leads to more time to find, debug, and fix the defect.
- ***Obscure unit tests***. Unit tests act as live documentation—they help understand the functionality of the underlying code. When unit tests are obscure, it becomes difficult to understand the unit test code as well as the production code for which the tests are written.
- ***Unit tests with external dependencies***. Unit tests should execute independently—they should isolate external dependencies while testing the subject under test. Failing to isolate real dependencies leads to many problems such as slow unit tests, failing unit tests due to a wrong reason, and inability to execute the tests on a build machine. For example, consider a case where a unit test directly accesses a database instead of isolating that dependency. The requirement to connect to the database and fetch the data will slow down the unit test. Further, the test may fail even though the underlying functionality is correct due to external reasons such as network failure.
- ***Improper asserts in unit tests***. There are many situations where wrong or non-optimal usage of asserts results in test debt. For instance, many a times where a "number obsessed" team adopts a bad practice wherein unit tests are written without necessary asserts to boost the coverage. Similarly, assert conditions are repeated that do not add any value to the test. Further, in many

[1]Note that the categories in this classification are neither mutually exclusive nor jointly exhaustive.

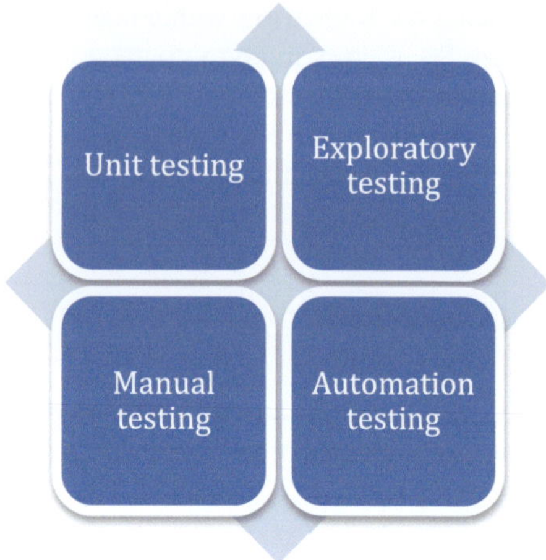

Fig. 2 Classification of test debt

situations, the provided assert conditions are wrong—in that case, it takes more time to debug the code to figure out the cause of the failure. Yet another case of improper asserts occurs when the test conditions are expected to be checked manually instead of providing them in assert statements. Such tests require manual intervention and effort to check the condition(s) every time the test is run that negate the benefit of having automated tests.

2.2 *Exploratory Testing*

The exploratory testing (or freeform testing) approach does not rely on a formal test case definition. In this form of testing, testers come up with tests and run them based on their intuition and knowledge instead of formally designing test cases [12]. Practitioners find exploratory testing attractive because of its creative, unconstrained, and cost-effective characteristics. Exploratory testing is also effective for certain kinds of tests. For example, security and penetration testing require considerable expertise and creativity from the test engineers to unearth security vulnerabilities.

However, overreliance on exploratory testing can undermine the overall effectiveness of the testing approach; some of the factors are:

- Exploratory testing typically increases effort and cost as the complexity of the application grows because it is harder to cover a large application with a few tests that the testers come up with rather than adopting a structured test management approach.
- Test managers have difficulty understanding and monitoring testing progress.

- Re-execution of tests is both difficult and expensive because the test cases are not documented. When test cases are not documented, the likelihood of some functionalities not tested is high, and hence the possibility of defects slipping to the field is also high.

Hence, we find that depending exclusively on exploratory testing contributes to test debt. However, when exploratory testing complements manual tests and automated tests, it is beneficial. In this context, following aspects can be identified that contribute to test debt.

- ***Inaccurate assessments***. The test-result assessment based on the missing Oracle in exploratory testing would result in additional rework due to more residual defects, directly affecting the maintenance costs.
- ***Inexperienced testers***. The main strength of exploratory testing derives from the experience of the testers and their domain knowledge. However, when inexperienced testers are employed to perform exploratory testing, the testing is ineffective. Due to suboptimal tester fitness and non-uniform test accuracy over the whole system, it leads to accumulation of residual defects.
- ***Poor documentation***. Because there is no documentation maintained when exploratory testing is performed, it jeopardizes knowledge management in the company resulting in higher maintenance costs. For instance, without documentation, testers new to the team will find it difficult to get up to speed and contribute effectively.

In addition, when proper logs of past testing activities are not maintained while performing exploratory testing, it can lead to other problems such as estimation errors in effort planning. Such improper effort estimation causes test debt because it encourages test engineers to take short-cuts in testing to meet deadlines.

2.3 Manual Testing

In manual testing, test engineers execute test cases by following the steps described in a test specification. Here is a list of common contributors to test debt that relates to manual testing.

- ***Limited test execution***. Many a time, the available resources and time to carry out complete and comprehensive testing is not sufficient. Hence, testers execute only a subset of tests for a release (i.e., a shortcut), thereby increasing the possibility of residual defects.
- ***Improper test design***. Manual testing is a time consuming and arduous process. It is the tester's responsibility to carry out all the documented tests and record their outcomes. A test case could be designed with lot of variations using different data combinations but then it is necessary for tester to execute all these combinations. Since execution of all combination of test cases is a laborious

process for testers, they restrict themselves to few happy scenarios for testing taking a short cut in test design. This increases the risk of residual defects in the System Under Test (SUT).

- ***Missing test reviews***. Test reviews help improve quality of test cases and also help in finding the problems earlier. Due to time pressures or process not mandating test reviews, testers take short cut by skipping test review. Missing test review could delay finding defects or increase maintenance of test cases.

2.4 Automation Testing

Automation testing involves executing pre-scripted tests to execute and check the results without manual intervention. Depending purely on manual testing and lacking automation itself would contribute towards test debt. For instance, it is essential to perform regression testing whenever the code is changed, and automated tests are key to effective regression testing. When automated tests are missing, it makes regression testing harder and increases the chances of residual defects, thereby decreasing the quality of the software and increasing the cost of fixing defects when they are reported later by the customers.

Here we consider contributors to test debt in projects where automated testing is performed. Since unit tests are developer centric, we consider debt related to unit tests different from automation test debt here.

- ***Inappropriate or inadequate infrastructure***. Tests being conducted in infrastructure not resembling customers infrastructure could lead to unpredictable behavior of the software resulting in reduced confidence on the system. Relevant hardware components or devices must be available for testing. For instance, in the context of embedded systems where both hardware and software components are involved, testing is often performed using emulators or simulators. Though it simplifies testing, skipping tests on actual hardware devices or components is a form of short-cut and it contributes to test debt.
- ***Lack of coding standards***. Automated tests need to adhere to common coding standard. When a common coding standard is not adopted and followed, it increases the effort to maintain the test code.
- ***Unfixed (broken) tests***. Not fixing broken tests or tests not being updated when functionality changes reduces confidence on the tests, decreases quality and increases maintenance effort.
- ***Using record and replay***. It is easy to use record and replay tools for testing Graphical User Interface (GUI) applications. However, using record and replay has numerous drawbacks—for example, even a slight change in a UI can result in the need for updating the test(s). Better alternatives may be available to use instead of record and replay tools. Hence, when record and replay tools are used

for GUI testing because they are easy to create (a form of shortcut), in the long run, it increases maintenance effort and contributes to test debt.

3 Managing Test Debt

Incurring debt is inevitable given the realities of software development projects that operate under schedule pressures with resource and cost constraints. The most important aspect in managing test debt (and technical debt in general) is to track the debt and periodically repay it to keep it under control. The key is to adopt a diligent and pragmatic approach towards evaluating and repaying technical debt.

However it is important to note that sometimes incurring debt is beneficial since it empowers the team to meet its goals. For example, it may not be possible to complete addressing all the findings from a test review because of the immediate release deadline that must be met. Not addressing the findings from test review incurs test debt, but it is acceptable so long as a plan is in place to address the findings in the upcoming releases and the plan is properly tracked and executed to repay the debt.

Another important aspect towards managing test debt is to "prevent" accumulation of test debt. Prevention can be achieved by increasing awareness in the development and test teams on test debt. Introducing relevant processes can also help stop accumulation of debt. For example, the focus given for "clean coding" [13] practices for software code needs to be given for test code as well. When existing processes are not adhered to or followed well, they can be strengthened again. For example, regular test reviews provide feedback loop to improve the quality of tests.

3.1 General Process for Repaying Test Debt

The key to manage test debt is to "repay" test debt. Often, software projects have accumulated considerable amount of test debt over a period of time. In such situations, it is important to evaluate the extent of test debt and plan to repay it. In real-world projects, it is not feasible or practical to stop writing new tests and focus exclusively on repaying test debt. Hence, a balanced and pragmatic approach is needed to repay debt that takes into account the project and organizational dynamics. With this in mind, we discuss steps for repaying debt based on our experience in industrial projects.

1. *Quantify the test debt, get buy-in from management, and perform large-scale repayment*
   When the debt is really high, it may be necessary for a project team to repay test debt by spending many months of dedicated effort to repay debt. Buy-in (i.e., acceptance and support) from management is often necessary for repaying test debt in these cases. To get buy-in, it is important to quantify the extent of test debt in the projects. For software code bases, numerous tools (such as SonarQube tool [14]) are already available which quantify technical debt for code and design debt dimensions. Since test debt is an emerging topic, tools are yet to emerge that quantify test debt. However, quantification can be still performed manually using simple tools such as excel sheets. Note that the term "technical debt" itself was coined as a metaphor to communicate the costs of taking short-cuts in development projects; hence it is important to be specific about the kinds of test debt incurred in the project and quantify (to the extent possible) to present the extent of debt to the management. Presentation to the management on test debt for getting their buy-in for repaying the debt should also include a plan for the team on how they will be repaying the debt. Such a plan would mainly cover time schedule, resources required and how they would be coordinated.
2. *Pay technical debt periodically*
   In projects where the debt has not reached critical levels, it is better for the team to repay debt in the form of small installments—equivalent of Equated Monthly Installments (EMIs) in financial domain—is desirable and practical. In this approach, make test debt repayment as part of every release along with the regular work of the test team. The Boy's scout rule "always leave the campground cleaner than you found it" is applicable for software development as well: "check-in the code better than you checked-out". Whenever developers or test engineers touch the tests, encourage them to improve the tests by refactoring them before checking them in.
3. *Avoid increasing test debt*
   Strategic approaches to managing test debt needs to be taken to avoid increasing test debts debt in future. To give an example, consider introducing Test Driven Development (TDD) [15] to the development team (in a team that hasn't adopted it yet). In TDD, the developer first writes an (initially failing) automated test case that defines a desired improvement or new function, then produces the minimum amount of code to pass that test, and finally refactors the new code and test to acceptable standards. In TDD, code for automated unit tests always gets written, executed and refactored. Hence, introducing TDD in a development team is a strategic approach towards managing test debt.

3.2 Strategic Approaches Towards Managing Test Debt

For pragmatic management of technical debt, tactical approaches such as refactoring test code may not suffice. Strategic approaches involving introducing new practices or adopting relevant practices is important. In this section, let us discuss a few important engineering practices that can play a significant role in managing test debt.

3.2.1 Applying Effective Coding Practices for Test Code

It is sometimes surprising to observe that software teams don't give the same importance to test code as given to application code shipped to customers. Towards rectifying this situation, it is imperative to follow best practices for managing code and design debt for reduction of test debt.

- *Pair programming*. Pair programming involves two programmers collaborating by working side-by-side to develop software; one developer is a driver who is involved in writing code and another one is a navigator who observes and helps the driver [16]. Benefits of pair programming include reduced defects, enhanced technical skills and improves team communication [17]. Test engineers, writing tests using pair programming, is also likely to bring in the same benefits.
- *Clean coding*. As Robert C Martin observes [13]: "Even bad code can function. But if code isn't clean, it can bring a development organization to its knees. Every year, countless hours and significant resources are lost because of poorly written code." Clean coding practices are now getting widely practiced with the spirit of craftsmanship for developing high-quality software [18]. In the same vein, even bad tests can find defects, but it is important that the tests are clean. Hence, it is important to adopt *clean testing* practices as a strategic approach for managing test debt.
- *Refactoring*. The traditional meaning of refactoring—"behavior preserving program transformations" [19]—is applicable for code in unit tests as well. Refactoring for unit tests is defined as transformation(s) of unit test code that do not add or remove test cases, and make test code better understandable or maintainable [11]. Refactoring is best performed as part of the development process, i.e., whenever the code is being changed (for fixing or enhancing code). In the same vein, test refactoring is best performed whenever the tests are being touched. The longer the refactoring is delayed, and more code is written to use the current form, it results in piling up of debt. Hence, refactoring unit tests and automated tests is the key practice for keeping test debt under control.

An important observation here is that these are not isolated practices—they are interdependent to each other. For instance, adopting clean coding practices requires making changes to the code for addressing static analysis violations, adopting and then ensuring that the whole codebase is written following a common coding style, etc. In this process, the developer making changes also naturally finds improvement opportunities in designs and performs refactoring. Any code refactoring needs corresponding unit tests to pass. When unit tests fail, typically, it is the code that is the culprit and that needs to be fixed. However, in many cases, the tests may also need to be refactored. Hence, as a result of introducing clean coding practices, code refactoring as well as test refactoring could take place [20].

Improved collaboration between developers and test engineers can also help avoiding/incurring test debt. For instance, "war room" approach in Agile methods and XP [21]. In this approach, developers and testers sit in the same room and develop the software. Any issues found by the tester are immediately discussed with developers and fixes are made immediately. This approach allows developers to cross-check with the testers as well. Such close collaboration serves well to managing test debt.

3.2.2 Applying Effective Testing Practices

There are many testing practices that software teams can adopt for strategic management of test debt.

- *Ten minute builds*. Quick and complete builds are important for reducing the turn-around time for making changes to the code in developer's machine to moving it to production. A rule of thumb is that it should take only 10 min to build, test, and deploy the software: starting from compiling the code, running tests, configuring the machines (registry setting, copying the files etc.), firing up the processes, etc. should take only 10 min or so and not hours [22]. In legacy software, it is not uncommon to see builds and deployment processes taking many hours—speeding it up to the order of 10 min is feasible and can be achieved by adopting relevant practices (such as incremental compilation) [23].
- *Scriptless test automation*. Conventional approach for GUI testing is to use record-and-playback tools: they are quick to create but are harder to maintain (e.g., even small changes in GUI can easily break the tests). GUI testing can be automated with scripts, but is effort intensive and requires considerably skilled test engineers; further, they also have maintenance problems. An emerging approach is to perform scriptless automation in which tests are composed from objects and actions through a GUI. The scriptless approach are less susceptible to changes in the GUI, technology agnostic, accepting late changes, and are relatively easier to maintain [24].

4 Case Studies

In this section, we discuss two case studies relating to test debt in industrial projects. These two case studies are derived from consultancy experience of one of the co-authors in this chapter. In documenting these case studies we have taken care not to mention any organisational or project-specific details.

4.1 Case Study I

This case study concerns a software product consisting of around 30 KLOC (Kilo Lines of Code) in a very large organization. It is a C++ software that has evolved over a period of eight years. Though the size of the code is small, it exposes more than a thousand Application Programming Interface (API) functions and can also be invoked through command line. It is a critical software used by a large number of teams within the organization. The application has to work on different platforms (Linux flavours, Windows and Mac). Team consists of 4 developers and 3 testers. A release is made every three to six months, with around 500–600 lines changed or added in every release.

The development process used was Waterfall model. Only manual testing was performed in this software when the feature additions or bug fixes were made. There was no unit testing performed in this software. No tools or processes were used by either developers or the testers to improve code quality. Test machines and test infrastructure used were obsolete. There were delays to get new hardware for testing because of arduous and time-consuming procurement processes followed in the organization.

The main problem that concerned the project team and the management was that the internal users were not satisfied because of the numerous defects in the software. Developers in the team found it difficult to make changes to the software (though it is a small software) because there was no automated regression testing in place. Since the test team had to perform testing manually, it took considerable amount time and effort from them.

One could observe that lack of automated (unit and functional) tests contribute to test debt. Further, processes in place were not conducive to creating high-quality software given the criticality of the software, or instance, lack of tools and processes for improving code quality.

The project team, with buy-in from management, took steps to address test debt in this project. The first step was to automate the tests and the team used GTest and Perl scripts. Automation was not challenging because the software was exposed as an API. Unit tests were implemented using CppUnit. For the command-line version, the team introduced tools such as Cucumber and Robot. The team also introduced BullsEye code coverage tool to track the extent of code covered by tests.

These changes took more than six months of time to implement and it delayed the next release of the software. But the changes were got beneficial to the product as the number of defects that slipped in, reduced. Further, the development team had more confidence in making changes because of the safety net available in the form of automated regression tests. The test team had a difficult time learning how to automate, but after automation was complete, they had more time to focus on finding hard-to-find defects using manual testing. The team now plans to integrate code quality tools as part of builds and plans to modernize its test infrastructure.

This case study illustrates how test debt can have significant impact even in a small code base. Tool driven approach and getting buy-in were strategic to the success in repaying test debt. However, repaying debt completely is impractical in real-world projects given the time, cost, and resource constraints faced by team as this case study illustrates.

4.2 Case Study II

This is the case of a critical financial product developed in a mid-sized organization. The project team consisted of 72 engineers (including development and test engineers). The code was written mainly in C++ with some parts written in Java. The size of the product was around 3 Million Lines of Code (MLOC). It was supported in multiple platforms. The original product was desktop based; recently the product is available for access from web as well. The product was a legacy software of 14 years old. The product followed Waterfall development approach.

Because of various factors, the organization owning the product decided to move the ownership of the product from country X to country Y. The plan was that the team in country Y will take up complete ownership of the product after 6 months time. During the 6 months time, the knowledge transfer should happen from the team in country X to the new team in country Y. After the ownership transfer, the new team had one year time for the next release that would include development of new features and fixing major defects reported by customers.

The new project team in country Y had to assess the product for taking ownership of the product. One of the key steps they planned was to conduct different kinds of assessments on the product including test assessments. Gathering inputs from the team in country X and analysing the project artifacts, the assessment team summarized its findings.

The product mainly consisted of manual test cases. There were approximately 1000 integration tests and 2000 system test cases. The tests were complex; sometimes few of the tests involved executing hundreds of steps. The test assessment team found considerable duplication between the tests across these three kinds of tests. Further, the steps within the tests were also duplicated.

The assessment team found that though there were automated tests available, they were not in usable condition. Recent changes for supporting the software in new platforms and exposing the software through a web interface has rendered the automation tests unusable.

The other challenges that the new project team in country Y faced was the lack of support from the project team responsible for knowledge transfer in country X. For this reason, the new project team wasn't able to get the automated tests working. Further, when the product was transitioned to country Y, the team size was reduced from 72 to 33 staff members. The new team consisted of only 6 test engineers.

Before the end of knowledge transfer for the product, the new test team performed a detailed analysis for estimating effort required for testing. Since it was a critical product all tests needed to be executed and it was estimated it would require 2 calendar years for complete execution of the tests. However, given the requirement that the next release should happen in one year timeframe, the team had to find a way to optimize the tests.

Since many of the tests in integration and system tests were the same due to duplication, the new test team removed redundant tests after careful analysis. Further, within the tests, many test steps were duplicated. Hence the team involved the product owner for analysis and review. Careful optimization of the tests reduced the integration tests from 1000 to 400 and system tests from 2000 to 700. The product owner gave considerable inputs on the correctness of these condensed tests.

To further optimize the time required for testing, the test team used a risk-based approach to prioritize the tests based on the criticality of the functionality. With that the tests were assigned with one of three priority levels. Priority 1 and 2 were considered mandatory for test execution whereas priority 3 tests could be executed opportunistically. After completing knowledge transfer (which took 6 months time), it took 8 months time to complete these steps to reach to a state where testing could actually begin.

Reflecting on this experience, we find that test debt accumulated over a decade had the impact of considerably slowing down the testing process. Further, huge amount of effort was unnecessarily wasted due to the accumulated debt. The specific kind of debt accumulated in this case relates to extensive test duplication. This shows that test debt is a key concern that software teams must care about.

Though automated tests were present in this software project, they were not in usable state. Whenever automated tests are broken due to changes in the software, it is important to get them working; hence, continuous maintenance and refactoring is important for automation testing.

In the upcoming release, the new test team is taking many steps for improving the test scenario such as bringing in smoke tests, introducing exploratory tests. The team also plans to get automated testing in place and improve the test review processes.

5 Future Directions

Technical debt as a metaphor has been gaining wider acceptance in academia and industry in the last decade. There are books [25], workshops [26], numerous research papers and conference talks covering technical debt. Some dimensions of technical debt—such as code and architecture debts—have received attention of software engineering community. Though there are some papers, blogs and articles that cover test debt, it is yet to gain wide attention from the community. In this section we outline some of the areas that need further exploration.

- The concept of "test smells" has been widely used for aspects that indicate test debt. Meszaros [27] differentiates between three kinds of test smells: *code smells* are bad smells that occur within the test code; *behaviour smells* affect the outcome of tests as they execute; *project smells* are indicators of the overall health of a project. Of these three kinds, code smells in unit testing has received considerable attention from the community (such as [11, 27–30]). However, other kinds of test smells such as behavioral smells and project smells haven't yet received much attention from the community.
- Mainstream Integrated Development Environments (IDEs) including Visual Studio, Eclipse, and IntelliJ IDEA support automated code refactoring. Though there are test smell detection tools available (such as TestLint [31]), we are not aware of any tools or IDEs that support automated refactoring for test smells.
- There are technical debt quantification tools that cover code and design dimensions (for example, SonarQube tool [14]). Tools that quantify test debt are yet to emerge.
- Test debt is an excellent metaphor to communicate the cost of short-cuts taken during the testing process to the management. In this context, test maturity assessment frameworks and methods (such as [32, 33]) can take test debt into consideration.

Given the importance of managing test debt in industrial projects, we hope that software testing community would address these areas in future.

References

1. G. Suryanarayana, G. Samarthyam, T. Sharma, *Refactoring for software design smells: managing technical debt* (Morgan Kaufmann/Elsevier, 2014)
2. W. Cunningham, *The WyCash Portfolio management system, experience report, OOPSLA '92* (1992)
3. I Gat, Opening statement on technical debt special issue. J. Inf. Technol. Manage., Cutter IT J. (2014)
4. Jim Highsmith, Zen and the Art of Software Quality, Agile2009 Conference, 2009
5. Z. Li, P. Avgeriou, P. Liang, A systematic mapping study on technical debt and its management. J. Syst. Softw. (2014)
6. R.L. Nord, I. Ozkaya, P. Kruchten, M. Gonzalez, *In search of a metric for managing architectural debt*. Joint 10th Working IEEE/IFIP Conference on Software Architecture (WICSA) and 6th European Conference on Software Architecture (ECSA), Helsinki, Finland, August 2012
7. A. Martini, J. Bosch, M. Chaudron, *Architecture technical debt: understanding causes and a qualitative model*. 40th EUROMICRO Conference on Software Engineering and Advanced Applications (SEAA) (2014)
8. K. Wiklund, S. Eldh, D. Sundmark, K. Lundqvist, *Technical debt in test automation*. IEEE Sixth International Conference on Software Testing, Verification and Validation (2013)
9. K. Pugh, The risks of acceptance test debt. Cutter IT J. (2010)
10. A. Qusef, G. Bavota, R. Oliveto, A.D. Lucia, D. Binkley, *Scotch: test-to-code traceability using slicing and conceptual coupling*. Proceedings of the 27th IEEE International Conference on Software Maintenance (2011), pp. 63–72

11. A.D. Leon, M.F. Moonen, A. Bergh, G. Kok, *Refactoring test code*. Technical report (CWI, Amsterdam, The Netherlands, 2001)
12. S.M.A. Shah, M. Torchiano, A. Vetro, M. Morisio, *Exploratory testing as a source of technical debt*", IT Prof. **16** (2014)
13. R.C. Martin, *Clean code: a handbook of agile software craftsmanship* (Prentice Hall, USA, 2009)
14. G. Campbell, Patroklos P. Papapetrou, *SonarQube in action* (Manning Publications Co., USA, 2013)
15. K. Beck, *Test-driven development: by example* (Addison-Wesley Professional, USA, 2003)
16. L. Williams, R.R. Kessler, *Pair programming illuminated* (Addison-Wesley Professional, USA, 2003)
17. A. Cockburn, L. Williams, *The costs and benefits of pair programming*. Extreme Programming Examined (2000)
18. S. Mancuso, *The software craftsman: professionalism, Pragmatism, Pride* (Prentice Hall, USA, 2014)
19. W.F. Opdyke, *Refactoring object-oriented frameworks*, Ph.D. thesis (University of Illinois at Urbana-Champaign, Illinois, 1992)
20. A. van Deursen, L. Moonen, *The video store revisited—thoughts on refactoring and testing*. Proceedings of International Conference on eXtreme Programming and Flexible Processes in Software Engineering (XP) (Alghero, Italy, 2002), pp. 71–76
21. O. Hazzan (ed.), *Agile processes in software engineering and extreme programming*. Proceedings of XP 2011 (Springer, Berlin, 2011)
22. K. Beck, C. Andres, *Extreme programming explained: embrace change*, 2nd edn. (Addison-Wesley, USA, 2004)
23. J. Shore, S. Warden, *The art of agile development* (O'Reilly Media, USA, 2007)
24. V. Moncompu, *Agile test automation: transition challenges and ways to overcome them*. Pacific NW Software Quality Conferences (2013)
25. Managing Software Debt, *Building for inevitable change* (Addison-Wesley Professional, Chris Sterling, 2010)
26. Workshop Series on Managing Techincal Debt, Carnegie Mellon—Software Engineering Institute, http://www.sei.cmu.edu/community/td2014/series/. Last accessed 29 Aug 2015
27. G. Meszaros, *xUnit test patterns: refactoring test code* (Addison-Wesley, USA, 2007)
28. B. Van Rompaey, et al., *On the detection of test smells: a metrics-based approach for general fixture and eager test*. IEEE Transac. Softw. Eng. (2007)
29. H. Neukirchen, M. Bisanz, Utilising code smells to detect quality problems in TTCN-3 test suites. Test. Softw. Commun. Syst. (Springer, Berlin, 2007)
30. B. Gabriele, et al., *An empirical analysis of the distribution of unit test smells and their impact on software maintenance*. 28th IEEE International Conference on Software Maintenance (ICSM, UK, 2012)
31. S. Reichhart, T. Girba, S. Ducasse, Rule-based assessment of test quality. J. Object Technol. **6** (9), 231–251 (2007)
32. I. Burnstein, A. Homyen, R. Grom C.R. Carlson, *A model to assess testing process maturity*. CROSSTALK (1998)
33. J. Andersin, *TPI–a model for test process improvement*. Seminar on Quality Models for Software Engineering (2004)

Agile Testing

Janakirama Raju Penmetsa

Abstract In recent times, software development has to be flexible and dynamic due to ever-changing customer needs and high competitive pressure. This competitive pressure increases the importance of Agile methods in software development and testing practices. Traditional testing methods treat development and testing as a two-team two-step process. The process discovers bugs in software at later stage of development. Further, the process frequently leads to an internal division between teams. Agile testing combines test and development teams around the principles of collaboration, transparency, flexibility, and retrospection. This testing enables the organization to be nimble about uncertain priorities and requirements. It helps to achieve higher quality in software products. This chapter with a brief introduction on Agile-based software engineering deals with Agile-based testing. Agile testing focuses on test-first approaches, continuous integration (CI), and build–test–release engineering practices. The chapter also explains advantages and disadvantages of Agile testing practices. The process is explained with an example.

Keywords Agile testing · Testing in parallel · Team collaboration · Continuous improvement · Test engineering

1 Introduction

In traditional testing, a two-team, two-step process is followed where development team builds softwarc to the state of perfection. Thereafter, testing team strives hard to find bugs in software and sends test report back to the development team. Only minimal collaboration happens between developers and testers through reviewing test case documents, design documents, requirement documents, etc. Sometimes, the reviews are ignored to accommodate change and critical timelines. This traditional process requires more time and money and often leads to an illusory divide

J.R. Penmetsa (✉)
International Game Technology (IGT), Seattle, WA, USA
e-mail: Janaki.Penmetsa@igt.com

H. Mohanty et al. (eds.), *Trends in Software Testing*,
DOI 10.1007/978-981-10-1415-4_2

Fig. 1 Cost of change curve

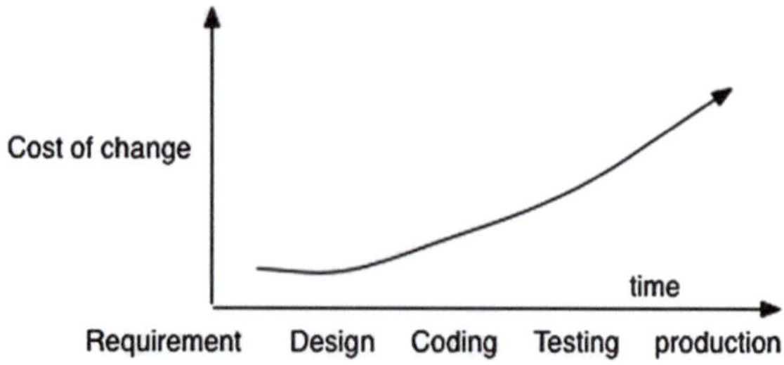

among developers and testers. The cost of change to fix a bug increases exponentially based on the time delay between the introduction of a bug and its finding. This phenomenon is described in Fig. 1.

Testing as close to development as possible is the key to Agile testing. Critics often call traditional methods as heavily regulated, regimented, and micro-managed. New Agile and light-weight methods evolved in mid-1990s to avoid shortcomings in traditional methods. Once Agile manifesto defined in 2001, its practice has come to prominence. The common aspect of all Agile methodologies is delivering software in iterations, keeping user priority in mind. Changes in requirements may change iteration priorities, but it is easy to reschedule the iterations as situation demands. The majority of successful Agile teams used the best development practices to come up with their own flavor of agility. Scrum and XP are two such popular Agile methods.

Agile testing means testing within the context of an Agile workflow. An Agile team is usually a cross-functional team, and Agile testing involves all of them in the process. Tester's particular expertise is used to embed quality into the deliverables at every iteration (Fig. 2).

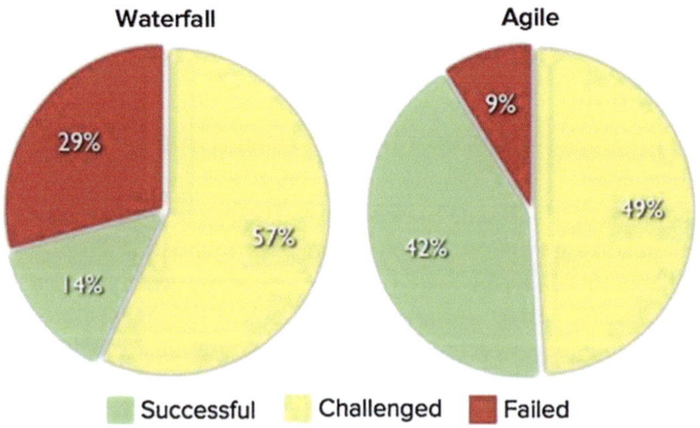

Fig. 2 As per CHAOS Manifesto 2012, a project is "Successful" means "Project is on budget and on time as initially specified". A project is "Challenged" means "Completed but over budget, over time". A project is "Failed" means "Project canceled at some point". *Source* The CHAOS manifesto, The Standish group, 2012

According to CHAOS Manifesto, Agile methods are about three times more successful than traditional methods as shown in Fig. 2 [1]. Keeping in this view, Agile testing has gained acceptance among practitioners.

In this chapter, Agile testing process is described in detail using a simple illustration. In Sect. 2, a short description on traditional testing process is described. Sections 3 and 4, respectively, present discussions on Agile-based software engineering and Agile-based testing. Section 5 presents an example explaining Agile testing process. Next section briefs on engineering process in Agile testing. Section 7 presents a brief analysis highlighting advantages and disadvantages in Agile testing. The chapter ends with a concluding remark in the next section.

2 Traditional Testing Practices

Traditional testing practices revolve around the testing center of excellence model [1]. A group of seasoned testers form as a separate testing team, who has the best customer focus, motivation, and intuition. Development team handles code delivery and fixing bugs.

Typically, after developers deliver code, testing team tries to find as many bugs as possible. Developers keep poor attention to quality while developing code. Consequences are harder and costly to fix.

Traditional testing teams keep costs low by using tools to document test cases and bugs and sometimes through labor outsourcing. However, this does not reduce systemic issues and shift costs back upstream into the development cycle. It results in higher levels of scrap and rework.

Preparing detailed test cases appear to help and optimize testing, but exacerbate the problem whenever requirements change. Change in requirements is almost unavoidable. Extensive efforts of testing activities slow down delivery. Even throwing a phalanx of testers is not very efficient. In reality, daily builds, functional and nonfunctional testing cadence overhead nullify any gains achieved due to specialized testing teams and practices.

Usually, testing is pushed to end of projects and gets squeezed. Unfortunately, when projects fall behind schedule, teams compress and sacrifice testing time to make up for delays in other processes. Thus, quality is always compromised.

Last-minute discovery of defects results in waste and high rates of rework. Longer it takes to provide feedback to developers, longer it takes to fix them. Developers take additional time to re-engage with the context of the code as they move on to the new project and new problems. The problem is worse if that last-minute bug is an architecture or design issue. A simple misunderstood fundamental requirement can cause havoc to timelines if discovered last minute.

Teams build up too much technical debt (also known as code debt or design debt). Technical debt is a metaphor referring to the possible future work of any system architecture, design, and development within a codebase [2]. Solution to these problems is to push testing to earlier phases of the development cycle.

In the mid-1990s, to overcome some of the above deficiencies, a collection of lightweight software development methods evolved. After Agile Manifesto published in 2001, these methods are referred to as Agile methods. Often loosely abbreviated as Agile, with a capital "A," next section describes Agile software practices in more detail.

3 Agile-based Software Engineering

The Agile manifesto, principles, and values of Agile software engineering are formed revealing better ways of producing software.

3.1 Agile Manifesto [3]

Agile Manifesto uncovers better ways of developing software by giving value to:

- **Individuals and interactions** over processes and tools,
- **Working software** over comprehensive documentation,
- **Customer collaboration** over contract negotiation,
- **Responding to change** over following a plan.

3.2 Agile Processes

Agile-based software engineering incorporates the principles mentioned just before. Several Agile processes such as Scrum, Kanban, Scrumban, Extreme Programming (XP), and Lean methods are in practice. Each one follows the Agile principles. The majority of successful Agile teams tune the processes with their own particular flavor of agility. Among those, this chapter describes two common processes Extreme Programming and Scrum.

3.3 Extreme Programming (XP)

XP is a software development methodology emphasizes teamwork and advocates frequent "releases" in short development cycles.

In short, take every observed effective team practice and push it to the extreme level. Good code review process pushed to the extreme of pair programming for instant feedback. Good software design is pushed to the extreme of relentless refactoring. Simplicity is pushed to the extreme of the simplest piece of software that could possibly work. Testing is pushed to the extreme of test-driven development and continuous integration (CI).

Managers, customers, developers, and testers are all equal participants in a collaborative team. XP enables teams to be highly productive by implementing a simple, yet effective self-organizing team environment [4]. XP improves a software deliverable in five essential ways; simplicity, communication, respect, courage, and feedback. Extreme Programmers regularly communicate with their fellow developers and customers. They keep their design clean and simple.

Testers start testing on day one and developers get feedback immediately. Team delivers the system as soon as possible to the customers and implements suggested changes. Every small success increases their respect for the individual contributions of team members. With this foundation, Extreme Programmers can courageously respond to changing technology and requirements [4].

3.4 *Scrum*

Scrum is a management framework for incremental or iterative product development utilizing one or more self-organizing, cross-functional teams of about seven people each. Within the Scrum framework, teams create their own process and adapt to it [5] (Fig. 3).

A product owner produces a prioritized wish list called as the product backlog. At the beginning of the iteration, the team takes a small chunk from the top of that backlog, and it is called as sprint backlog. The team then plans on how to implement those pieces in detail. The team has about two to three weeks time to complete its work. However, the team meets every day to assess its progress called as daily Scrum. Along the way, the Scrum master responsibility is to keep the team focused on its goal. The work is potentially shippable and demonstrated to the stakeholder.

Fig. 3 Scrum framework [5]

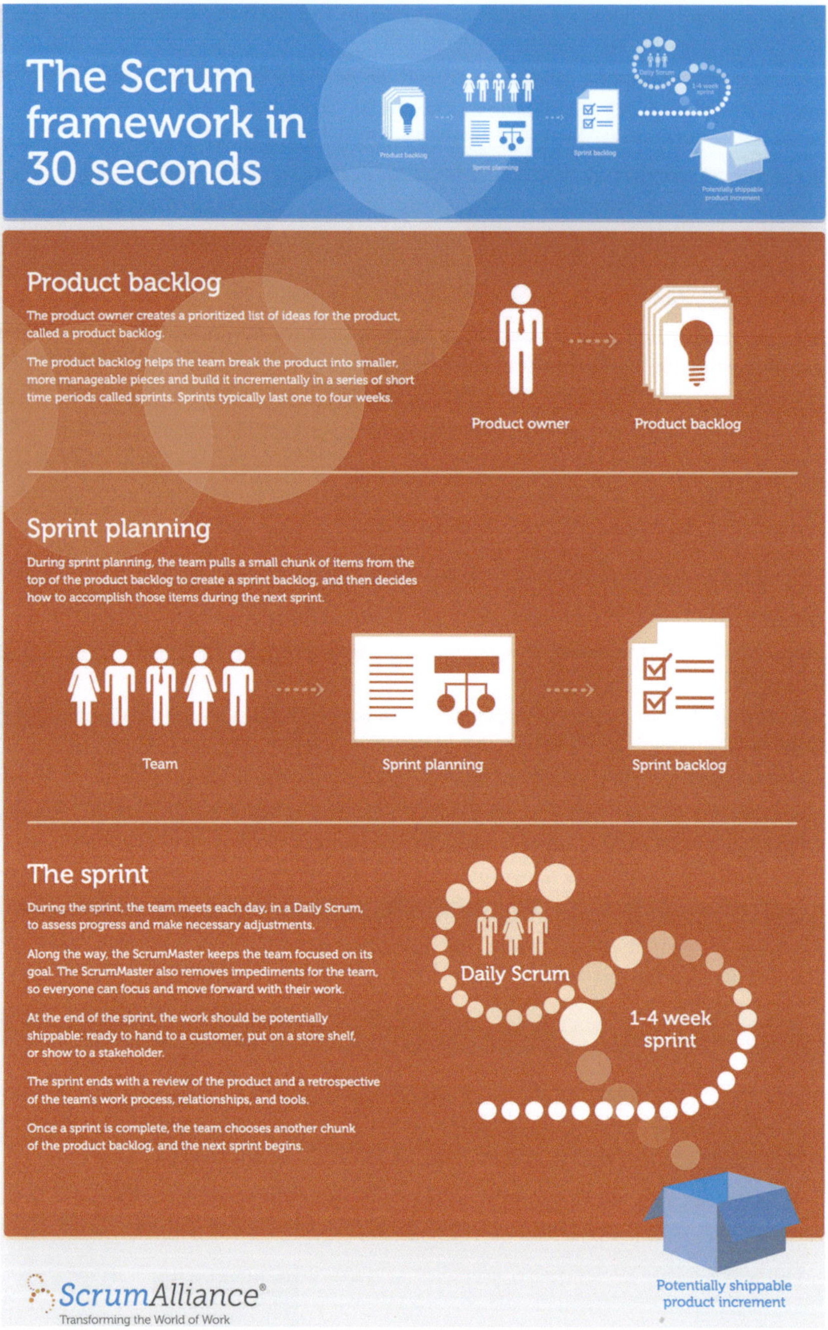

Fig. 4 Scrum process [6]

The sprint ends with a sprint retrospective and review. And the whole process repeats, with next chunk of work from the product backlog [5] (Fig. 4).

After understanding Agile processes in detail, it is time to explore Agile-based testing (at times referred as "Agile testing") and how it helps to achieve higher quality.

4 Agile-based Testing

Agile-based testing means testing within the context of an Agile process such as Scrum or XP. Agile suggests development and testing, two essential functions of software development, proceed concurrently. *Agile testing emphasizes on positive desire to implement a solution that passes the test, confirmation of user stories, whereas traditional testing methods focus on negative desire to break the solution, falsification of given solution.*

Development and testing teams are united based on Agile principles to perform Agile testing. Illusory divide between code breakers (testers) and code creators (developers) is reduced. The necessity of both development and testing roles is respected. The testing team works to provide feedback to developers as soon as possible. This integration implies both that developers cultivate skills of testers and testers understand the logic in development.

Applying Agile concepts to the testing and QA process results in a more efficient testing process. Pair programming is encouraged to provide instant feedback, and also test-driven methodologies come handy, where tests can be written before the actual development. Lack of automation is the principle barrier to provide instant feedback to developers. End-to-end automated testing process integrated into development process achieves continuous incremental delivery of features.

In short, Agile requires teamwork and constant collaboration among software development teams and stakeholders. Quality is "baked into" the product at every phase of its development through continuous feedback. Everyone in the team holds a vision of the final product.

Working software is preferred over documentation and produced in small iterations frequently. However, every iteration delivered is in line with demands of the final system. Every iteration has to be fully integrated and carefully tested as a final production release.

In Agile testing, there is no strict adherence to requirement documents and checklists. The goal is always to do what is necessary to complete customer's requests. Documentation is often replaced with constant collaboration through in-person meetings and automated tests. Only essential and minimal documentation is maintained. Unified self-organizing project teams work closely together to achieve a high-quality software product. The notion of separate testing team disappears entirely. In some cases, it might be necessary to do separate release candidate testing for not to find problems or bugs. It is just performed for verification, trail audit completion, and regulatory compliance.

The continuous internal dialogue between team provides better-working relationships between testers and developers. Constant collaboration with end users improves responsiveness to ever-changing project requirements.

The retrospective meeting is an integral part of Agile process and is done after every iteration for continuous improvement. This meeting is an integral part of Agile process. These meetings reflect self-organization and regular adaptation of the team.

5 Illustration

Let us consider a use case to design a small messaging service to understand the process better. This new service sends a message to the user if the user is online. Otherwise, it would persist until the user comes online. Assume Scrum process is being followed to implement this use case.

Developers and testers discuss customer requirements and identify the tests that need to be designed for acceptance criteria. A lot of questions are raised by the testing team during this discussion to drive clarity for the requirements (e.g., message persistence duration, the number of messages limits, and the priority of messages). Development teams focus on identifying challenges in solution design. These discussions usually happen in pre-iteration planning meetings (also referred as backlog grooming meetings). Also, the work is divided into several small user stories. User stories are one of the primary development artifacts for Scrum project teams. A user story is a simplified representation of a software feature from a customer point of view. It describes what customer wants in terms of acceptance criteria. Also, it explains why the customer needs that feature that helps in the understanding of business value.

For example, following user stories are identified for the “small messaging service”:

The division of the user stories is completely at the discretion of the team and business owner and what they think appropriate to divide the work. Prioritization of the stories is also done during the meeting based on the effort and priority from the business owner.

After prioritization of stories is completed, and a backlog is prepared, a sprint/iteration can be started by taking a small chunk from the top of the backlog. In the sprint, developers and testers work in parallel. Every day, the team meets and discusses what has to be done that day and what testing has to be done to verify the work. Quality is thus embedded into working software delivered by the team with constant collaboration, negotiating better designs for easy testability.

Testers participate from day one and insist on writing code that is testable. This emphasis often leads to use design patterns extensively and achieves better code layout, architecture, and quality. Design pattern means a reusable solution to a commonly occurring problem within a given context in software design.

In the present illustration, as testers and developers approach first user story in Table 1, they have to agree on a standard API or interface to write their code or test, to start their work in parallel as in interface-based programming pattern. Teams agree upon interface "FetchUndeliveredUserMessageResource" that contains "fetchUndeliveredUserMessages" method, to accept "userId" as the parameter as described in Table 2. Testers start writing their integration tests, and developers start their implementation in parallel.

Interface-based programming is claimed to increase the modularity of the application and hence its maintainability. However, to guarantee low coupling or high cohesion, we need to follow a couple of guiding principles. First one is single responsibility principle. It suggests that every class should have the responsibility for a single functionality or behavior, and the class should encapsulate that implementation of that responsibility. This principle keeps tedious bug prone code contained. The second one is interface segregation principle. It suggests that no client should be forced to depend on methods that do not use. In that cases, interfaces are from the third party, using adapter pattern that makes the code well organized. The basic idea of adapter pattern is changing one interface to the desired interface.

Table 1 User stories

1. As a user, I want to receive all persisted messages at login so that I can know all the pending messages. *Acceptance Criteria:* *- all messages received when the user offline has to be received by the user when he comes online* *2. As a user, I want to persist messages when I am not online so that those messages are not lost.* *Acceptance Criteria:* *- store user message, if not delivered.* *3. As a user, I want to receive messages when I am online so that I can be aware of events happening around me.* *Acceptance Criteria:* *- when online, receive notifications immediately as they happen.* *4. As a user, I want to receive messages in a priority order, so that important messages are received first.* *Acceptance Criteria:* *- sort messages based on the priority and deliver the important one first.* *- Higher the priority.. higher the importance.* *etc.,*

Table 2 *FetchUndeliveredUserMessageResource* interface

```
@Path("/api")                                                             // HTTP base URI Java Annotation
public interface FetchUndeliveredUserMessageResource {
        @POST                                                             // Supports both POST and GET
        @GET
        @Path("/user/{userId}/getundeliveredmessages")                    // URI for this resource
        @Consumes("application/json")                                     // Content-Type is application/json
        void fetchUndeliveredUserMessages(@PathParam("userId") String userId );  // userId is the parameter
}
```

In the present illustration, as testers and developers approach second user story in the Table 1, the team identifies the need to separate database retrieval so that it can be tested independently and defines "PersistMessage" interface to access data from the database as described in Table 3. This repository design pattern comes handy to separate business logic, and data model, and retrieval. This pattern centralizes the data logic or Web service access logic and provides a substitution point for the unit tests and makes it flexible.

In the present illustration, as testers and developers approach third user story in Table 1, "SendMessageResource" interface has been identified as described in Table 4. As the testing team writes tests for these user stories, they would design them to run in isolation and test each story independently and in isolation. It reduces test complexity and forces development to reduce complex dependencies between components. As much as possible, no test is made dependent on another test to run.

Also, to reduce interdependencies among different components and classes, dependency injection pattern is used.

Dependency injection is a software design pattern that implements inversion of control for software libraries. That means component delegates to external code (the injector) the responsibility of providing its dependencies. The component is not allowed to call the injector system. This separation makes clients independent and easier to write a unit test using stubs or mock objects. Stubs and mock objects are used to simulate other objects that are not in the test. This ease of testing is the first benefit noticed when using dependency injection. Better testable code often produces simple and efficient architected system.

Table 3 *PersistMessage* interface

```
public interface PersistMessage {
    public boolean persist(Message msg);
    public boolean delete(Message msg);
}
```

Table 4 *SendMessageResource* interface

```
@Path("/api")
public interface SendMessageResource {

        @POST
        @Path("/api/type/{userId}/sendmsg")
        @Consumes("application/json")
        void sendMessageToUser(Map<String, Object> data);
}
```

When the sprint is in progress, if new work is identified, or priorities changed by business, product backlog is groomed as appropriate.

A retrospective meeting is required to talk freely about what good practices has to be continued and what practices need to be stopped or improved. This meeting happens at the end of the sprint. As the team progress during the sprint and integrate their work and do build and test, fixed as often they can, it necessitates better test engineering practices.

6 Engineering of Agile Testing

Agile methods do not provide actual solutions but do a pretty good job of surfacing problems early. The motto is to "catch issues fast and nip them in bud."

Agile testing team has to test every iteration carefully as if it is a production release and integrate it into the final product. As a result of repeated integration tests, testing team needs to design and implement test infrastructure and rapid development and release tools. The success of Agile testing depends on the team's ability to automate test environment creation and automatic release candidate validation and report test metrics.

6.1 Continuous Integration

As teams deliver fully functional features every iteration in Agile practice, they need to work ways to reduce build, deployment, and running test overhead. CI is a practice that requires developers to integrate code several times a day. Each time, a developer checks in code, which is verified by an automated build, allowing teams to detect problems early. CI allows automated tests to grow, live, and thrive. By integrating regularly, errors can be detected quickly and easily.

6.2 Automated Build System

Automated build system enables the team to do CI. It has become a cornerstone of Agile development.

Automated build system reduces the cost associated with bug fixing exponentially and improves quality and architecture. There have been several sophisticated continuous build systems available such as Jenkins [7], Hudson [8], Team Foundation Server [9], and Apache Continuum [10].

Build systems need to be optimized for performance, ease of use by developers, incremental execution of tests, and software language agnostic.

In the above illustration, the team used Jenkins as their automated build system. Every time, a developer checks in code, or tester checks in a test in which a new build is made, and all the unit and integration tests are executed. If any issues are found, a build failure email is sent to the team. Jenkins makes the deployment artifact as an Redhat package manager (RPM). These RPMs are deployed to the team environment by deployment scripts automatically so that testers can perform post-deployment verification as needed. Next section does a comparative study of Agile testing and its advantages and disadvantages.

7 Agile Testing: An Analysis

Let us briefly review the distinction between Agile and Spiral models, as there is apparent similarity between the two. Then, an analysis is made showing both advantage and disadvantage in Agile testing [11].

7.1 Comparison of Agile and Spiral Model

The common aspect of Agile process and Spiral model is iterative development and differ in most of the other aspects. In case of Spiral model, length of iteration is not specified, and an iteration can run for months or years. But in Agile model, length must be small and usually, it is about 2–3 weeks. In Spiral model, iterations are often well planned out in advance and executed in order of risk. In Agile process, focus is on one iteration at a time, and priority of user stories drives the execution order instead of risk. Within each iteration, Spiral model follows traditional waterfall methodology. Agile recognizes the fact that people build technology and focuses "test-first" approaches, its collaboration, and interactions within the team.

7.2 *Advantages of Agile Testing*

Knowledge transfer happens naturally between developers and testers as they work together with constant feedback on each iteration. Junior team members benefit from active feedback, and testers perceive the job environment as comfortable. Resilience is built into the teams and nourishes team atmosphere.

Small teams with excellent communication require small amounts of documentation. This collaboration facilitates learning and understanding for each other. There is no practical way to comprehend or measure this particular advantage but produces an awesome environment and team.

Requirements volatility in projects is reduced because of small projects and timelines. The wastage of unused work (requirements documented, unused components implemented, etc.,) is reduced as the customer provides feedback on developed features and helps prioritize and regularly groom the product backlog. Customers appreciate active participation in projects.

Small manageable tasks and CI helps process control, transparency, and increase in quality. Very few stories are in progress at any point in time and handled in priority order, and this reduces stress on teams. The risk associated with unknown challenges is well managed. Frequent feedback makes problems and successes transparent and provides high incentives for developers to deliver high quality.

As bugs are found close to the development, they can be fixed with very limited extra overhead and thus increase actual time spent on designing and developing new features (Fig. 5).

The quality of work life improves as the social job environment is trustful, peaceful, and responsible.

7.3 *Disadvantages of Agile Testing*

As the focus on intra-team communication increases too much, it sometimes leads to inter-team communication issues, and "us" versus "them" attitude between the teams. Teamwork is critical to the success of Agile testing, and it is essential that team members are willing to work together. Personalities of the team members play a big role in Agile practices.

Agile methods have lots of challenges with scaling them to large groups and organizations. Coming up the division of teams and projects is often not trivial needs lot of commitment and patience. Generating the priority list of all the projects, broken up into small stories, is a herculean task and so also to maintain.

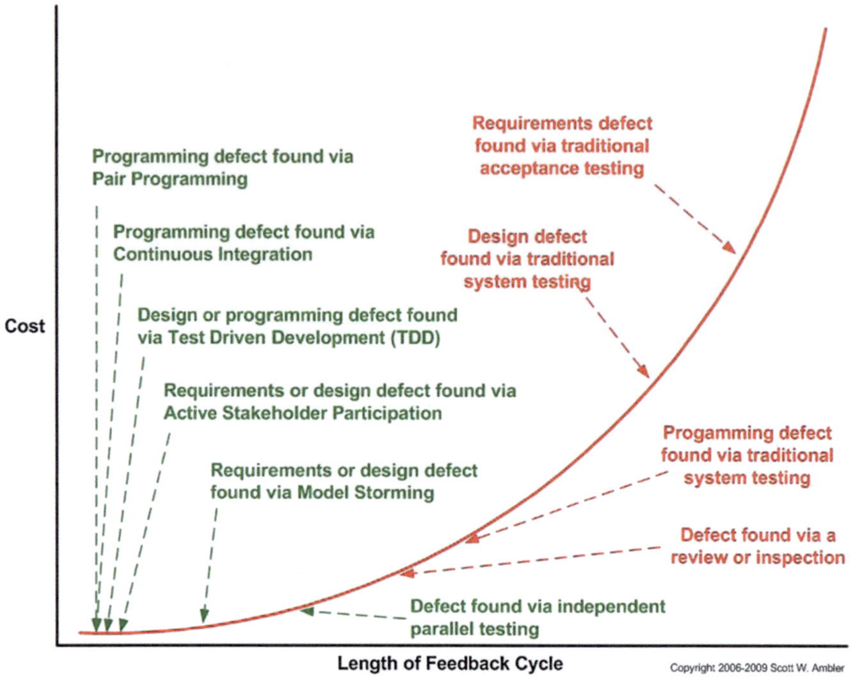

Fig. 5 Cost of feedback cycle [12]

Implementation starts very early in the process, and sometimes, enough time is not spent on design and architecture. Poor design choices might lead to rework. Sometimes, dependencies buried inside implementation details are hard to identify, often causes impediments to the team. Many people would be required to manage increased number of builds, releases, and environments.

8 Conclusion

Agile testing combines test and development teams around the principles of collaboration, flexibility, simplicity, transparency, and retrospection. Its focus is on positive desire to implement a solution that passes the test, rather than negative desire to break a solution. Agile practices do not provide actual solutions but do a pretty good job of surfacing them early. Developers and testers are empowered to work toward solving problems.

Agile has already won the battle for mainstream acceptance, and its general precepts are going to remain viable for some time. As empowerment of people is key to good governance so is for software development. Agile approaches empower a team possibly replacing command control structures in organizations with more democratic practices. So, the success of this process is well predicted. Hence, both academia and industry are currently showing increasing interest in Agile testing. This chapter with an intention to highlight this upcoming approach from industry perspective has briefly described both Agile software development process and Agile testing approach. The approach has been illustrated with a case study. We aim the chapter will be useful to the beginners interested in this nascent area of software testing.

References

1. http://www.computerweekly.com/feature/Why-agile-development-races-ahead-of-traditional-testing
2. https://en.wikipedia.org/wiki/Technical_debt
3. http://www.agilemanifesto.org/
4. http://www.extremeprogramming.org/
5. https://www.scrumalliance.org/why-scrum
6. https://www.scrumalliance.org/scrum/media/ScrumAllianceMedia/Files%20and%20PDFs/Why%20Scrum/ScrumAlliance-30SecondsFramework-HighRes.pdf
7. https://jenkins-ci.org/
8. http://hudson-ci.org/
9. https://www.visualstudio.com/en-us/products/tfs-overview-vs.aspx
10. https://continuum.apache.org/
11. http://www.sciencedirect.com/science/article/pii/S0164121209000855
12. http://www.ambysoft.com/essays/agileTesting.html
13. http://www.agilemanifesto.org/principles.html
14. https://www.mountaingoatsoftware.com/blog/agile-succeeds-three-times-more-often-than-waterfall

Security Testing

Faisal Anwer, Mohd. Nazir and Khurram Mustafa

Abstract Security issues in application domain are often easily exploited by attackers. In today's scenario, the number of vulnerabilities is enormously increasing leading to serious security threats in future. Security threats may arise due to distributed and heterogeneous nature of applications, their multilingual and multimedia features, interactive and responsive behavior, ever evolving third-party products and the rapidly changing versions. Recent report reveals the fact that security threats surrounding the financial applications have increased dramatically and they continue to evolve. Hence, it is imperative to make the software applications more secure and reliable. Security testing is an effort to reveal those vulnerabilities, which may violate state integrity, input validity and logic correctness along with the angle of attack vectors that exploit these vulnerabilities. Further, security testing minimizes risk of security breach and ensures confidentiality, integrity and availability of customer transactions. This chapter illustrates significance and relevance of security testing in present context. It will help students, researchers, industry practitioners and security experts. Further, it will give some of the possible directions of research in the area.

Keywords Application security testing · Secure software development life cycle · Phase embedded security testing

---

F. Anwer
Department of Computer Science, Aligarh Muslim University, Aligarh, India
e-mail: faisalanwer.cs@amu.ac.in

Mohd. Nazir · K. Mustafa (✉)
Department of Computer Science, Jamia Millia Islamia (A Central University), Jamia Nagar, New Delhi, India
e-mail: kmustafa@jmi.ac.in

Mohd. Nazir
e-mail: mnazir@jmi.ac.in

H. Mohanty et al. (eds.), *Trends in Software Testing*,
DOI 10.1007/978-981-10-1415-4_3

1 Introduction

Now a days, industry is producing software for almost every domain that is generally complex and huge in size. Majority of the software are being used openly in the prevailing distributed computing environment, leading to a range of security issues in application domain. These issues are exploited by attackers easily and so the number of attack incidences on software applications is growing rapidly. As the number of vulnerabilities is enormously increasing, there are possible serious threats and challenges in future that can cause a severe setback to applications. Security threats also arises due to distributed and heterogeneous nature of applications, their multilingual and multimedia features, interactive and responsive behavior, ever evolving third-party products and the rapidly changing versions. Recent reports reveal that security threats surrounding the financial and banking applications have increased dramatically and they continue to evolve.

It is imperative now to develop the software applications, with high security attributes to ensure reliability. In the view of the prevailing circumstances, security is one of the major concerns for many software systems and applications. Industry practitioners assume that installed security perimeter such as antivirus, firewall are secure enough and they believe them as effective measures for dealing with the security aspect of an application. With the number of security products growing up, there is still lack of attention in resolving security issues in online banking systems, payment gateways, insurance covers etc. IDG survey of IT and security executives report that nearly 63 % of applications remain untested for critical security vulnerabilities [1].

Security is a process, and not a product. It cannot be stapled or incorporated to an application during finish time, as an afterthought. Security testing defines tests for security specific and sensitive requirements of software. Security requirements are functional as well as non-functional and thus require a different way of testing compared to those applied for primary functional requirements. Security testing is an effort to reveal those vulnerabilities, which may violate state integrity, input validity and logic correctness along with the angle of attack vectors that exploit these vulnerabilities. Further, security testing minimizes the risk of security breach and ensures confidentiality, integrity and availability of customer transactions. As a tester, it is harder to devise exhaustive anti-security inputs and hardest to prove whether application is secure enough against these infinite set of input. Hence, it is necessary prerequisite to understand security challenges and debatable issues in order to comprehend existing methodologies and techniques available for security testing.

Security is not one-time approach too, but a series of activities performed throughout SDLC starting from very first to the last phase, which leads to SSDLC (secure software development life cycle). Security testing is the most important aspect in the SSDLC as it is an important means to expose the serious security related issues of the applications. Researchers and practitioners advocate security testing as a series of activities that are performed throughout system development process. If security testing is taken care of in early stages of SDLC, security issues can be tackled easily and at a lower cost. Hence, it would be instrumentally effective as well as efficient, if security testing is embedded with each phase of SSDLC. Security testing should start from requirement phase where security requirements are tested against security goals. At design phase, security testing may be carried out using threat model and abuse case model [2]. Secure coding in coding phase may be validated against static analysis methods/tools [3] such as Flawfinder, FindBugs. Security testing at testing phase is carried out as functional security testing as well as risk-based security testing [3]. Security testing at this stage is carried on as attacker's perspective. Finally, production environment needs to be properly tested which ensures that the application is protected against the real-time attacks in the live environment. All these activities form the approach namely as phase embedded security testing (PEST).

This chapter illustrates significance and relevance of security testing in present context and organized as follows: Sect. 2 elucidates the current security challenges faced by applications. Section 3 highlights the significance of security testing. Section 4 gives an overview of SSDLC while Sect. 5 discusses the security issues and the related concerns. The approaches of security testing are discussed in Sect. 6, and phase embedded security testing approach (PEST) is explained in Sect. 7. Section 8 discusses industry practices while Sect. 9 highlights industry requirement and future trend. Finally, Sect. 10 concludes the chapter with some of the important directions of research in this area.

2 Current Security Challenges

Security is an essential mechanism that restricts attackers from exploiting the software. Modern applications are facing several security challenges that if not properly tackled may inject serious flaws in an application. A number of open security issues reportedly being perceived as challenges by research community. This paper discuss security challenges that exhibit significant impact on the applications such as software complexity, rise of third-party code and dynamic security policies reported by various researchers and practitioners [4–7].

2.1 Software Complexity

Software, both as a process and the product, is becoming more and more complex due to prevailing practices in industry. Primitive software was having limited functionalities with limited users but due to advancement in technologies and increase in dependencies on software, software is not only highly complex but also huge in size and virtually expandable operational environment. Complex software has more lines of code and has more interactions among modules and with environment. It is harder for users to understand complex applications and thereby makes it difficult to test, which may ultimately in untested portions leading to greater possibilities of subtle bugs in the system. More the bugs in the system, more is the possibilities of introducing security vulnerabilities. Experts argue that complexity of software makes them more vulnerable [8].

Department of Defence [9] report states "The enormous functionality and complexity of IT makes it easy to exploit and hard to defend, resulting in a target that can be expected to be exploited by sophisticated nation-state adversaries." This shows that complexity is actually enemy of security. Complex system may have loopholes that may be exploited and difficult to diagnose because of complexity. Mccabe Software argues that complexity may leave backdoors and attackers may implant Trojan code, which is difficult to identify because of complexity [10]. This discussion raises an obvious question how to avoid or minimize complexity that is now days one of the debatable issues at various forums.

In his famous book entitled as "No silver bullet," Books argues that the complexity of software is an essential property, not an accidental one [11]. We cannot avoid complexity of software, but it can manage complexity to minimize its consequences as the complexity is realized to be the result of several elements prominently including problem domain complexity, difficulty of development process and its management, the flexibility that is possible through software and the problems of characterizing the behavior of discrete systems [12].

2.2 *Third-Party Code*

Today's applications are collection of different components build through different sources that includes in-house built, outsourced, open source, commercially built and others. Due to market pressure to produce software quickly and at low cost, software industries are using significant portion of code developed by third party. The presence of third-party code in applications represents a serious security risk for companies, according to a study from security vendor Veracode. Bugs in third-party library may lead host application as a whole to become vulnerable [13]. Security vulnerabilities in third-party code expose serious concerns since these vulnerabilities can affect a large number of applications.

There are several instances where the companies faced severe security threats due to third-party code. Recently, twitter came across security flaw of third-party code which saw a cross-site scripting flaw exploited on its site. The third-party code activates a JavaScript function called "onmouseover", which can activate pop-up box. This flaw could also be exploited to redirect a user to an infected Web site [14].

It is obvious that third-party code produces significant security concerns, but this does not mean that third-party code should not be used at all. The real concern is whether security measures have been implemented by third parties and whether they have been properly tested independently and with applications. According to the report of coverity [15], companies are using major part of software code from multiple third parties and these codes are not tested with the same intention and rigor as internally developed software. So the conclusion is that if third-party code is being used, it must be properly tested for security flaws.

2.3 Dynamic Security Policies

Information security policies specify a set of policies adopted by an organization to protect its information agreed upon with its management [16]. It clearly describes different parties with their respective part of information at respective levels. Information security policy document includes scope of the policy, classification of the information, management goal of secure handling of information at each class and others.

In traditional systems, these security policies are static in nature that is programmers can only specify security policies at compile time. However, modern systems interact with external environment where security policies cannot be known well in advance and applied. Thus, security policies are dynamic in nature for such cases like for instance deciding authority of users depending on type of message received through external environment. Therefore, a mechanism is essentially needed that can allow security critical decisions at runtime based on dynamic observations of the environment [14].

3 Significance of Security Testing

It is an important activity to uncover vulnerabilities in the software, which is also instrumental for minimizing the risk of security breach and ensuring confidentiality, integrity and availability of customer transactions. Security testing can no longer be perceived as an afterthought and stapled or incorporated at the last minute. It is now a full fledge activity. Researchers frequently advocated that it should be integrated with development life cycle and it must strictly be followed. Unfortunately, it is rarely being practiced in the industry. According to Trustwave global security report [17], 98 % tested applications were found vulnerable that clearly indicates the low level of seriousness in software industry toward security testing. A properly executed security testing with suitable technique would provide benefits to software industries, their clients as well as end users leading confidence building in the product. Following sections discuss the importance of security testing.

3.1 *Software Industry Perspective*

Industry generally presumes that security testing will not return on investment. They consider that they have completed functional testing and environment is fitted with firewall and antivirus, so there is no need to invest extra cost on security testing. But the fact is something different. A recent study reveals that three out of four applications are at risk of attack and large number of these attacks are performed on applications and firewalls and SSL cannot do anything in this case [18]. Actually in long run, industry will get following benefits through security testing.

3.1.1 Preserved Brand Image

A well-tested application goes through less downtime and enhances the brand image of industry. It will add value to the brand and indirectly will attract other client as well. Kirk Herath, chief privacy officer, assistant vice president and associate general counsel at Nationwide Insurance in Columbus, Ohio says "Any breach has the tendency to dampen greatly whatever you are spending around your brand" [19]. On the other side, brand damage can have serious concern for companies. They may lose trust of existing clients and new clients will be reluctant to use their products.

3.1.2 Reduced Time to Market

An untested or improperly tested applications need to fix the error at later stages which may need to redesign the application. This will eventually increase the time to market since redesigning an application needs to implement, test and deploy those new changes. Impact of these new changes also needs to be analyzed. These all efforts will result in late delivery of the product. Security testing can reveal several security issues early in the development which may be fixed with low cost and effort.

3.1.3 Lower Development Costs

A properly and thoroughly tested applications will face less failure; otherwise, cost to fix the application at later stages will be high and subsequently development cost will be higher in these cases. For example, fixing a requirement error found after an application has been deployed costs 10–100 times compare to 10 times when error found at testing stage [20]. Similarly, fixing a design error found after an application has been deployed costs 25–100 times compared to 15 times when error found at testing stage.

3.2 *Client Perspective*

There is no doubt that client is the ultimate and direct beneficiary of well-tested quality product. A client itself could be an industry such as retail, banking and insurance. Untested software may be exploited by the attackers that will have direct influence on client. They may lose important data and may be victim of financial loss. Therefore, clients are most concerned regarding the products and vendors who are supplying these. A well-tested product will have great significance for client including as follows.

3.2.1 Attack Resistant Product

An untested or improperly tested application will be victim of several attacks and will provide safe heaven for the attackers. A properly tested application will rarely face data breach, denial of service (DoS) and other security issues. During security testing phase, application will be tested for wide security vulnerabilities. It is also important that industry should involve security experts and effective security testing process to produce attack resistant product. Client will have confidence on their product, and their product will not be a victim of an attack.

3.2.2 Better Quality Software

Obviously, incorporating security on the product will add quality to the product. Security testing insures whether security aspects have been added to the product or not. A properly security tested product will face minimum service downtime and hence will improve quality of the product. On the other hand, an untested or improperly tested product will miss quality aspect of the product.

3.2.3 Minimizes Extra Cost

A small flaw in any part of the application can hamper performance of the critical services. They may need other hardware or software to compensate for this loss which in turn needs more investment. A single security flaw in an application may cost millions to the customer. A study carried out by Ponemon reports that security violation sometimes costs the businesses at an average of $7.2 million dollars per breach [21]. This shows a huge cost is associated with a single security failure. In some cases, industry may need to pay millions of dollars compensation.

3.3 End-user Perspective

End users are those who will be benefited with industries IT environment such as customers of an online banking. If the software is vulnerable to attack, then personal data, credential and others may be stolen by an attacker. End user may face interrupted services and they may be victim of financial lose also. So it is very important that end users are provided with properly protected and well-tested applications. Following are the significance of well-tested application in context of end-user perspective.

3.3.1 Uninterrupted Service

An insecure product may be victim of DoS attack which may completely block applications. End users will be unable to access the application because of this type of attack. A more sophisticated attack, distributed denial of services (DDoS), may be used by attackers. Well-known companies such as Twitter and Facebook in past have faced Dos attack [22]. Security testing ensures that applications should not be exploited by these types of attack, and at very early stage, the applications are protected from these attacks.

3.3.2 Minimizes Chance of Loss of Personal Data and Credentials

Attackers are very much interested in users credential as well personal data. They may use these data for their ill intentions. According to a recent article of CBS news [23], personal data of over 21 million individuals were stolen in a widespread data breach. It is very important for an application to guard against these data breaches. Application should be properly tested and protected for these types of breaches.

4 Secure Software Development Life Cycle

Security is not a one-time approach after implementation. However, it includes a series of activities performed throughout SDLC starting from very first to the last phase that leads to SSDLC. Security should be perceived as process, and not the product. Therefore, it can no longer be considered as an afterthought that to be stapled or incorporated to an application at the last minute. The industry that apply security throughout SDLC renders better results compare to those, which do after development.

Forrester Consulting conducted a survey study of 150 industries in 2010 to understand application security current practices and its key trends and market directions [24]. In this survey, they found that the industry that practices SSDLC

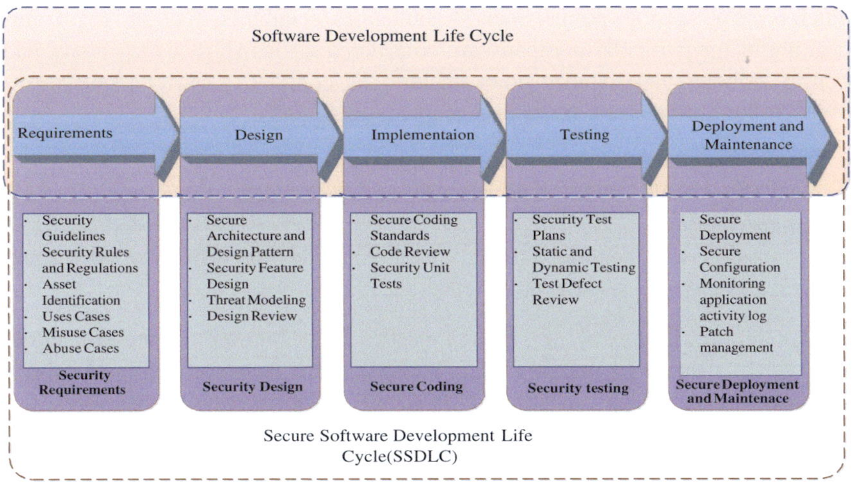

Fig. 1 Secure software development life cycle

specifically showed better ROI results compared to overall population. Another study [18] reports that fixing security vulnerability found when a product is live, around 6.5 times more costly than fixing those at early phase of SDLC. Security experts have proposed several methods to incorporate security in SDLC such as Microsoft's Security Development Lifecycle (SDL) [25], McGraw's Touchpoints [26] and OWASP's Comprehensive Lightweight Application Security Process (CLASP) [27]. Figure 1 depicts phases and set of task at each phase of SSDLC. Although phases are depicted here as traditional waterfall model, most organizations follow an iterative approach these days.

SSDLC incorporate security from very early in the development life cycle. It starts from requirement phase where security requirements are established and continue to design phase where secure architecture and design patters are used. Next comes implementation phase where secure coding standards are adopted, followed by testing phase where functional and non-functional security issues are tested and at last deployment and maintenance phase where software is deployed in secure environment. These phases in SSDLC are called security requirements, secure design, secure coding, security testing and secure deployment and maintenance. We briefly discuss each phases of SSDLC as follows.

4.1 Security Requirement

At very first company-wide security policies are established that ensure role-based permission, access level controls, password control and others. Requirement specification are thoroughly examined to decide where security can be incorporated which is called security points. Security guidelines, standards, rules and regulations meaningful for particular company are considered at this stage which along with security policies and security point decides the complete security requirements. Some of the sample security requirements are as follows:

- Application should only be used by legitimate users.
- Each user should have a valid account, and their levels of privilege should be properly defined.
- Application send confidential data to different stake holders over the communication network, so encryption technique should be applied.

Experts have proposed several approaches to accommodate security during requirement phase such as [28, 29]. A comparative study on different security requirement engineering methods can be found in Fabian et al. paper [30].

4.2 Secure Design

At design stage, security experts must identify all possible attacks [31] and design the system accordingly. These threats can be systematically identified through one of the popular technique called threat modeling [32]. It helps the designer to develop mitigation techniques for the possible threats and guides them to concentrate on part of the system at risk. Other technique to model software design is Unified Markup Language (UML) diagrams, which helps to visualize the system and include activities, individual components and their interaction, interaction between entities and others. UMLsec [33] an extension of UML for secure system development has been proposed in the literature.

4.3 Secure Coding

Application security vulnerability can be broadly categorized into design-level vulnerabilities and implementation-level vulnerabilities [34]. A design-level vulnerability occurs due to flaw in design such as using unsuitable cryptography. Implementation-level vulnerabilities exist due to flaw in coding such as improper use of error handling, not restricting unwanted inputs and so on. These flaws could lead to vulnerabilities such as denial of services, buffer overflow. During secure coding phase, secure coding practices are adopted to restrict implementation-level vulnerabilities. SEI CERT of CMU has provided a set of secure coding standards [35] for languages such as C, C++, Java and others. These standards can be applied by the developer to make the coding secure.

4.4 Security Testing

Security testing is an important phase in the SSDLC since it helps to improve software security. This should not be considered as an optional/secondary component of functional testing but should be treated as a full fledge activity. Security testing includes testing of functional security aspect such as testing of access control, authentications as well as application specific risk-based testing such as denial of services, buffer overflows. Several testing techniques have been applied in security testing such as random testing, search-based testing, symbolic execution and others, but random testing is most popular since it is easy to follow and discovers a wide range of security issues. We have discussed security testing techniques in detail in Sect. 6.

4.5 Secure Deployment and Maintenance

After successfully applying security at each phase the product should be installed in secure environment. Software and its environment need to be monitored continuously, and in case security issues are found in the product or environment, a decision should be made whether patch is required or not [34]. Software and its environment also need to update periodically so that they may not be affected by the security vulnerabilities. Attackers used to exploit outdated or older version of the software since chances of vulnerabilities are high in these cases.

5 Security Issues and Related Concerns

Insecure design and insecure implementation of an application lead to security issues that may be exploited by an attacker. A security practitioner should know these security issues to properly test and protect a system. Open source vulnerability database (OSVDB) [36] reveals that applications are encountering a significant amount of these issues periodically. Figure 2 shows occurrences of these vulnerabilities in each quarter. Among these, XSS (Cross-site scripting) is a major issue occurring these days. Here, in this section, we briefly describe security issues which are included in OSVDB.

5.1 Cross-Site Scripting (XSS)

This is a typical kind of loophole through which malicious scripts are injected through client-side script that access confidential data, content of html page may be

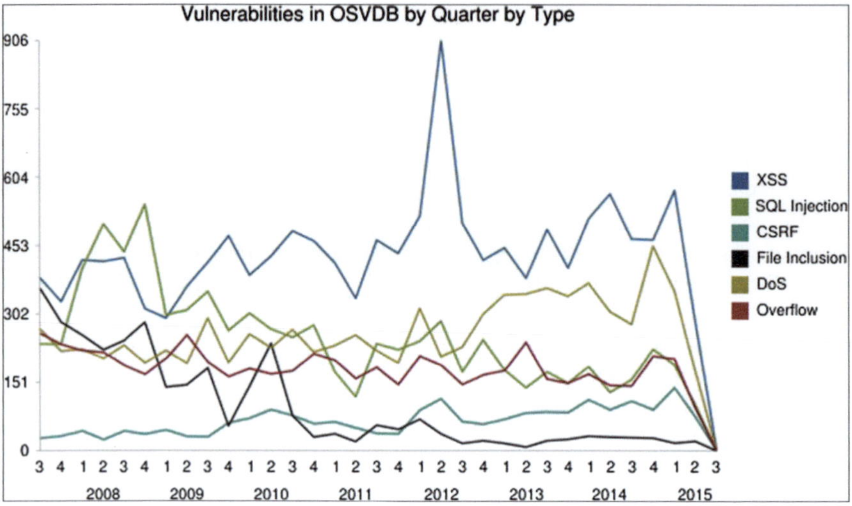

Fig. 2 Vulnerabilities in OSVDB by quarter by type [36]

rewritten by that script. This vulnerability is considered as one of the most popular hacking techniques to exploit Web-based applications. An attacker who exploits XSS might access cookies, session tokens and even they can execute malicious code into other users system [37]. In past, several popular Web sites were affected by XSS, prominently among these is recent instance of twitter that is exploited through this vulnerability [38]. This enables someone to open third-party Web sites in users browser just hovering mouse over a link.

5.2 SQL Injection (SQLI)

SQL injection is a vulnerability in which an attacker gained unauthorized access to the database using malicious code that includes SQL queries. Web applications face serious threat due to this vulnerability, and as per data of OSVDB, a large no of applications are affected by this vulnerability [36]. It results from supplying inadequate user inputs that use to construct database statements. Attacker exploits by inserting malicious script/code into an instance of SQL server/database which eventually attempts to fetch any database information. SQLI vulnerability allows attackers to disclose information on the system, or makes the data unavailable, and in extreme case administrators of the database server [39].

5.3 *Cross-Site Request Forgery (CSRF)*

Cross-site request forgery (CSRF) is a vulnerability in which unwanted actions are transmitted from an authentic user on a trusted site. This attack could lead to a fund transfer, changing of a password or purchasing an item on behalf of an authentic user. Unlike XSS in which malicious codes are injected through infected Web site, in CSRF, malicious codes are injected to a trusted Web site. A recent study [40] shows that Web sites of several well-known companies are vulnerable to CSRF. An example of CSRF vulnerability was once reported by Web site of ING DIRECT in which an attacker could open an additional account of a registered user and subsequently could transfer funds to his account [41].

5.4 *Denial of Services (DoS)*

Denial of services (DoS) attack, that targets service to make unavailable to intended clients, has shown serious threat to the Internet security [42]. DoS can result from various reasons such as application crash, data destruction, resource depletion like memory, CPU, bandwidth and disk space depletion [43]. DoS attacks were carried out with only little resources and thus caused a serious threat to organizations. A more sophisticated form of DoS is distributed DoS (DDoS) where several systems may be distributed across globe is used to carry on an attack.

5.5 *Buffer Overflow (BOF)*

Buffer overflow is very common and easy to exploit vulnerability. This vulnerability can be exploited to crash the application and in extreme case, an application can control the program. In a classic buffer overflow vulnerability, an attacker enters large-sized data into a relatively small-sized stack buffer, which results in overwriting of call stack including the function's return pointer [44]. In a more sophisticated attack, function's return pointer is overwritten with address of malicious code. Both Web server and application server can be exploited by BOF.

5.6 *File Inclusion*

File inclusion vulnerability is a type of vulnerability, which permits an attacker to attach a file on the Web server usually through a script. This occurs due to improper validation of user inputs. Hackers use two categories of file inclusion: remote and local file inclusion (RFI/LFI) attacks. The attack leads to sensitive information disclosure, DoS and execution of malicious code on the Web server/client side [45]. According to Imperva report, RFI was ranked as one of most prevalent Web application attacks performed by hackers in 2011 [46].

6 Security Testing Approaches

Security testing is an important activity to reveal those vulnerabilities which if exploited may harm the system. It adds quality to the product and brand value to the company. Day-by-day increase in exploitation of software and the use of sophisticated technologies are supporting attackers to succeed very much in their ill intention. So, it is essential to properly test the software for security vulnerabilities to protect it from attackers.

Researchers have proposed several security testing approaches in the literature to expose security issues. Security testing can be broadly categorized into static and dynamic depending on complexity of application and type of vulnerabilities to be found. Static testing is carried without executing the software such as code reviews, static analysis while dynamic testing is performed by executing the software. Here, in this section, we discuss these broad categories of security testing approaches. We include case studies of java programs each in category of static security testing and dynamic security testing. These programs are tested with tool FindBugs [47], a static analysis tool and Symbolic PathFinder [48], a concolic executor. Although other testing techniques have been discussed here under these categories, we applied only static analysis tool and concolic execution tool to java programs.

6.1 Static Security Testing

Static security testing automatically detects security vulnerabilities in software without executing the programs. Some static testing methods use source code for analysis and others use object code. Static techniques can be performed to a very large program but it may suffer from generation of false positive, false negative or spurious warning. Here, we are discussing main categories of static security testing such as code review, model checking and symbolic execution. Model checking and symbolic execution come under static code analysis techniques.

6.1.1 Code Review

Code review involves testing an application by reviewing its code. Its purpose is to find and fix errors introduced in the initial stage of software development, and hence, overall quality of software will be improved. Code review can be done manually as well through automated tools. Manual code review is very time-consuming which needs line by line scan of code where as automated code review scans source code automatically to check whether predefined set of best practices have been applied or not.

6.1.2 Model Checking

Model checker automatically checks whether a given model of the system meets given system's specification. Specification may contain safety, security or other requirements that may cause the system to behave abnormal. Systems are generally modeled by finite-state machines that would act as an input for model checker along with collection of properties generally expressed as formulas of temporal logic [49]. The model checker checks whether properties hold or violated by the system.

Model checker works in following manner. Suppose M be a model, i.e., a state-transition graph and let p be the property in temporal logic. So the model checking is to find all states s such that M has property p at state s. An important tool Java Path Finder (JPF) [50] based on model checking is widely used research tool for various different execution modes and extensions. Several testing tools based on JPF such as Symbolic PathFinder, Exception Injector and others have been proposed in the literature. A whole list of such methods/tools can be found on the site of JPF.

6.1.3 Symbolic Execution

This technique generates test cases for every path by generating and solving path condition (PC) which is actually a constraint on input symbols. Each branch point of the program has its own path condition, and PC at higher level is the aggregation of current branch and branches at previous level. Classical symbolic execution internally contains two components:

- **Path condition generation**: Suppose a statement is given as *if(cond) then S1 else S2*. PC for the branch of this statement would be $PC \rightarrow PC \wedge cond$ (for true branch) and $PC \rightarrow PC \wedge \neg cond$ (for false branch).
- **Path condition solver**: Path condition solver or constraint solver is used for two purposes, first to check whether path is feasible or not, and second, to generate the inputs for the program that satisfies these constraints. Constraint solving is the main challenge of symbolic execution since the cost of constraint solving dominates everything else and the reason is that constraint solving is a NP-complete problem.

Classic symbolic execution provides sound foundation for dynamic testing such as *concolic testing*. Plenty of works based on *concolic testing* have been proposed in the literature. We have discussed Concolic testing in next section.

Table 1 Bugs found through Findbugs tool

Java package name	Line of code	No. of classes	Total bugs found
BareHTTP	717	6	14

6.1.4 Case Study

In order to demonstrate the actual working of testing tool, we consider a case study of a java package, namely BareHTTP [51]. We have applied popular static testing tool Findbugs [47], a static analysis tool for java code on BareHTTP package that produces following result mention in Table 1. Findbugs takes java programs as input and generates different types of bugs such as empty database passwords, JSP reflected XSS vulnerability and so on.

Every Bugs found through Findbugs do not come under security issues. A careful analysis of bug report generated for BareHTTP shows that some bugs are generated because IO stream object is not closed. This may result in a file descriptor leak, and in extreme case, applications may crash. Tool in discussion generates wide list of bugs, and they are categorized under bad practices, performance issue, dodgy code, malicious code vulnerabilities and so on. All bugs produced by Findbugs are not actual error, and it may generates several false positives.

6.2 Dynamic Security Testing

Dynamic security testing involves testing of programs for security in its running state. It can expose flaws or vulnerabilities that are too complicate for static analysis to reveal. Dynamic testing exhibits real errors, but it may miss certain errors due to missing certain paths. Dynamic testing involves using a variety of techniques in which most popular are fuzz testing, concolic testing, search-based testing. Each one is valuable from certain perspective such as test effectiveness, test case generation, vulnerabilities coverage. Each of these techniques is described next along with latest research in these directions.

6.2.1 Fuzz Testing

Fuzz testing (random testing) is type of testing in which a program is bombarded with test cases generated by another program [52]. The program that generates test cases is called fuzz generator, which generates random, invalid or unexpected data to test the program. Intention of fuzz testing is to crash a program to expose security holes in the program. Early fuzz testing techniques were simple and used to generate random numbers as input for the program. Current fuzz testing techniques are more sophisticated in nature and generate more structured input as test cases.

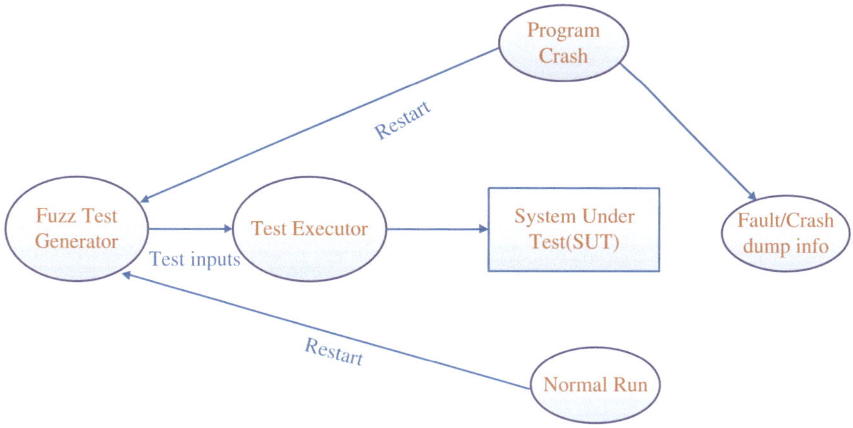

Fig. 3 Basic diagram of Fuzz Testing

Fuzz testing can be broadly divided into black box and white box fuzz testing. Black box fuzz testing generates input without knowledge of program internals on the other hand white box fuzz testing considers internals of the program also. As far as coverage of fuzz testing is concern, it can uncover several security issues like buffer overflow, memory leaks, DoS and so forth [53]. Fuzz testing process can be better depicted through Fig. 3. This figure shows that fuzz test generator repeatedly generates test inputs and the program gets executed with these inputs with the help of test executor. A log is also maintained to store the fault/crash data.

Researchers have proposed a number of fuzz testing techniques such as [54–58] to test programs for bugs that lead to program crash. Among these, popular methods are JCrasher [55] an automated testing tool that test for undeclared runtime exception which if executed will crash the program and CnC [56] an another random testing method that combines static testing and automatic test generation to find the program crash scenario.

6.2.2 Concolic Testing

Concolic testing is a type of testing that while running the program collects the path constraints along the executed paths and these collected constraints along the path are solved to obtain new inputs for alternative path [59]. This process is repeated till all the coverage criteria is covered. Basically, it is an extension of classic symbolic execution which dynamically generates inputs by solving the path constraints.

Concolic testing has been very successful in security testing since it executes each and every path in the program and make sure that program does not contain dark path (untested and neglected path). This testing technique automatically detects corner cases where programmer not handled exception properly or fails to allocate memory, which may lead to security vulnerabilities. Several methods/tools have been developed using concolic execution such as Cute and JCute [60], Symbolic JPF [48] and others. These methods are very effective in security testing of the programs since it tests all paths and hence avoid untested paths that may lead to security vulnerabilities.

6.2.3 Search-Based Security Testing

It is an application of meta-heuristic optimizing search technique to automate a testing task. It uses test adequacy criterion known as fitness function to automatically generate test data. So the goal is to satisfy fitness function starting with an initial input and continuously refining the inputs that maximize the satisfiability of fitness function. Although there are several search-based algorithm that have been applied in software security testing but genetic algorithm is most popular among these. Figure 4 presents an overview of main tasks of genetic algorithm.

Several methods [61–64] have been developed using search-based algorithms to test security issues such as buffer overflow, SQLI, program crashes and others. In one of these paper [63], program crash scenario due to divide by zero exception has been tested. In their method, the program is first transformed in such a way that the expression that leads to crash becomes condition. Using this condition, test data are generated using genetic algorithm.

6.2.4 Case Study

We have taken a case of program given in Listing 1 that generates uncaught divide by zero exception. Results of uncaught or improper handled exceptions can lead to severe security issues [63]. This may allow an attacker to cause a program crash which will finally lead to DoS. Several instances of divide by zero exceptions have been reported in vulnerability databases such as OSVDB [36], NVD [65]. As per database record, 9 such cases are in 2015, while 15 cases in 2014 [36]. For instance,

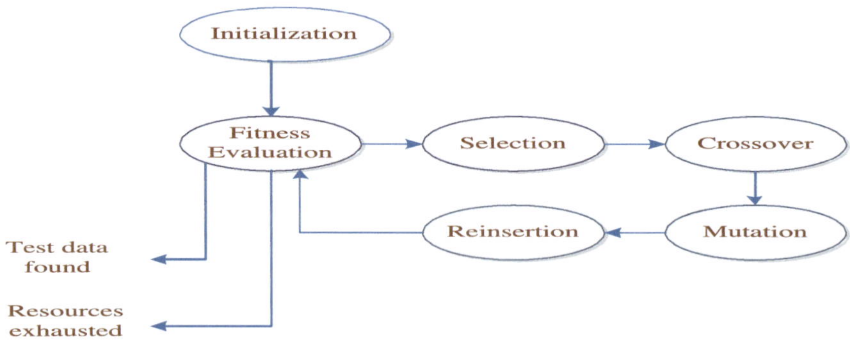

Fig. 4 Overview of the main tasks of a genetic algorithm

FFmpeg is reportedly suffering from divide by zero vulnerability in one of its function during handling of dimensions exploitable by a context-dependent attacker [66].

In the given program mentioned in Listing 1, divide by zero exception has not been handled at line numbers 7 and 12 that will eventually raise uncaught exception. This program has been tested using Symbolic Java PathFinder [48]. The symbolic Java PathFinder is developed as an extension of JPF that executes the program on symbolic inputs. Inputs are represented as constraints generated through conditions in the codes that are subsequently solved to generate test inputs.

```
1  class symexe{
2    int i,j,z;
3    void foo(int i, int j){
4      if (i>j)
5      {
6   System.out.println("i is greater than j");
7   z=i/(j-8);
8      }
9      else
10     {
11   System.out.println("j is greater than i");
12   z=j/(i-4);
13     }
14     if (z>5)
15   System.out.println("z is greater than limit");
16     else
17   System.out.println("z is less than the limit");
18  }
19
20    public static void main(String[] args){
21       symexe s = new symexe();
22       int p,k;
23       s.foo(12,6);
24    }
25  }
```

If we will constitute Symbolic tree of the above program, then it will generate six paths out of which two paths will generate uncaught exceptions. Symbolic PathFinder generates the constraints of these six paths, one of them is given below as:

(j 2 SYMINT / (i 1 SYMINT – CONST 4)) <= CONST 5 &&
(i 1 SYMINT – CONST 4) != CONST 0 && i 1 SYMINT <= j 2 SYMINT

These constraints for each path constraint are solved subsequently using different constraint solver embedded in Symbolic Java PathFinder and finally following results are produced:

Table 2 Bugs found through symbolic PathFinder

Java program name	LOC	No. of test cases generated	Total bugs found
DividebyzeroExample	25	7	2

```
[foo(54,11)]
[foo(92,51)]
[foo(86,–77)]
[(expected = java.lang.ArithmeticException.class), foo(21,8), ##EXCEPTION## "jav
a.lang.ArithmeticException: div by 0..."]
[foo(21,8)]
[foo(11,54)]
[foo(0,0)]
[(expected = java.lang.ArithmeticException.class), foo(4,49), ##EXCEPTION## "jav
a.lang.ArithmeticException: div by 0..."]
[foo(4,49)]
```

In normal setup, generating values for these cases are usually difficult since a program may have. Result of Symbolic PathFinder on the above program can be summarized in Table 2.

Here in the above test cases, two test cases foo(21,8) and foo(4,49) generate divide by zero exceptions. Other test cases would execute remaining paths other than the paths that generated the exceptions. This type of testing is very effective in the sense that it needs few number of test cases to find bugs in the program. We have taken a case of improper handling of "divide by zero" exception and tested it using a concolic testing tool. There are some other types of exploitable exceptions to crash a program, such as Null Pointer exception, that have not been taken in this study.

Security testing techniques discussed in Sect. 6 are further categorized. Table 3 gives brief description of each technique including attack types it covers like BOF, DoS and others mentioned in the table.

7 Phase Embedded Security Testing

In previous sections, we have discussed that security is not a onetime approach after implementation but a series of activities throughout development life cycle. Like security is phase-based activities, similar to this, we have proposed that security testing should also be carried out from the very first stage, i.e., requirement stage till deployment and maintenance phase.

Software security should be applied as two-layer approach. At very first, security need to carry out throughout development life cycle and in second-layer security testing should be embedded at each phase of SDLC. Figure 5 elaborates this concept. At very first phase, security requirements such as security rules, standards, policies and goals are collected and formalized. Test cases are created and analyzed

Table 3 Security testing techniques and attack type coverage

Testing techniques		Description	Attack type coverage
Static	Code review	• Audit source code for security vulnerabilities, manually or with tool • Capable of pinpointing security issues. Cost-effective but not suited for large application • Can be performed in parallel with coding	All kinds of attacks, depending on reviewers expertise/experience
	Model checking	• Verify if a model of system meets given security specification • Easy to regenerate test cases in response to changes • Strong coupling of the tests with requirements or design	SQLi, BOF, DoS, XSS
	Symbolic execution	• Tests programs through symbolic input and generate test cases through constraint solving • Wide coverage of control paths of program • Unable to solve complex constraints and ineffective in handling of external functions	SQLi, BOF, DoS, XSS
Dynamic	Fuzz testing	• Testing program using huge and random inputs • Easy to perform • Depth of coverage of control paths of program but miss wide path coverage	DoS
	Search-based security testing	• Testing using meta-heuristic optimizing search technique to automate security testing • Automatic generation of test cases • Works well if heuristic offers significance guidance. It may stuck in local optima	SQLi, BOF, DoS, XSS
	Concolic testing	• Concurrently executing one path and generating inputs for alternate path by negating collected path constraints along the executed paths • Wide coverage of control paths of program but may miss depth of coverage • Difficult to solve complex constraints and handling of external functions is ineffective	SQLi, BOF, DoS, XSS

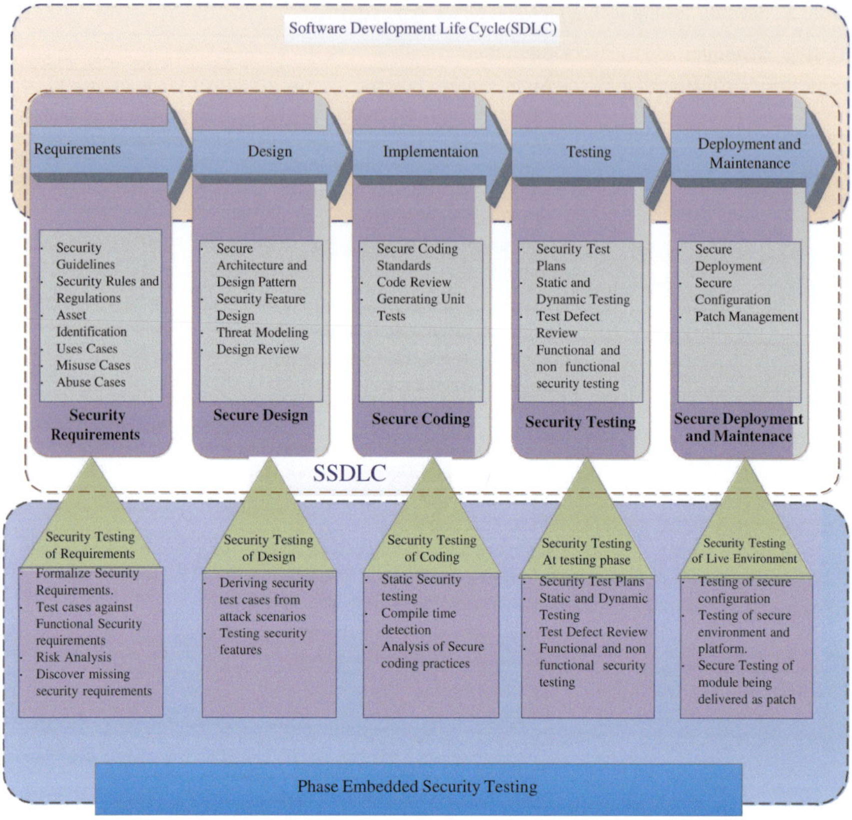

Fig. 5 Phase embedded security testing (PEST)

against these security requirements. In second phase, security test cases are derived from attack scenario and security features are tested. One such attack scenario could be finding ways of application crash. Model-based design verification is also helpful to test secure design. During security testing at coding phase, secure coding practices are analyzed, and static security testing is applied to detect security issues. At testing stage, functional and non-function security issues are tested using static or dynamic or combination of both. Finally, security testing of live environment is carried out through testing secure configuration, environment and platform being used. Modules delivered as patch should also be tested for security issues. Production environment needs to be properly tested as it will ensure that the application is protected against the real-time attacks in the live environment.

As a case study, we consider an organization that has formed software development team to build a Web-based application. Knowing the significance of security relating to application, the software development team made every effort to incorporate security in the application. Security experts were involved with the development team from initial requirement gathering phase till deployment and

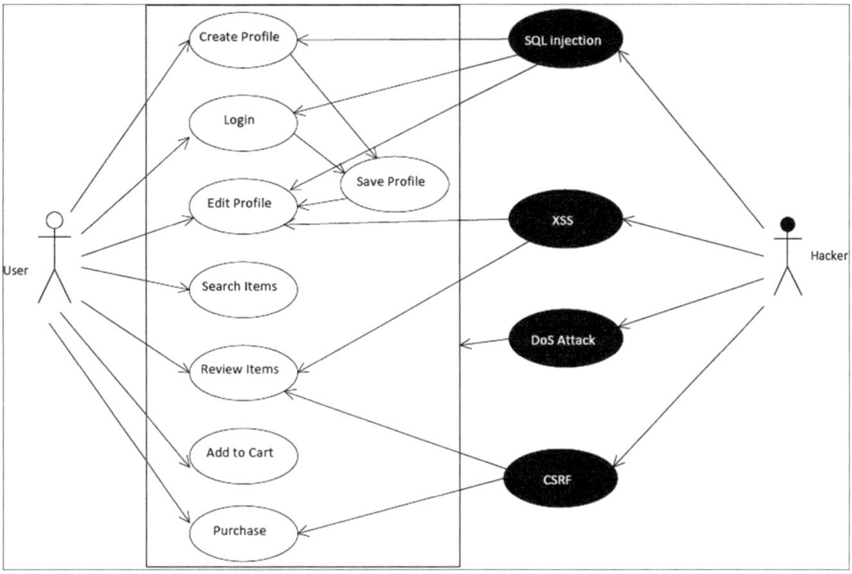

Fig. 6 Misuse case diagram for an online shopping Web application

maintenance. Security testing activities were conducted throughout the development cycle to ensure that security was built and thoroughly tested. Following activities performed during PESC for online shopping application at each phase.

- **Security Testing of Requirements**: First of all collected security requirements are modeled using misuse case. We have drawn misuse case of online shopping application as depicted in Fig. 6. Misuse case provides useful information to review security requirements by the domain and security experts. Below we have shown a misuse case scenario of improper input validation that may lead to XSS.
  - Hacker enters login information
  - System displays product menu
  - Hacker selects most sold product
  - System displays product details
  - Hacker enters review with a link of malicious Web site
  - Hacker logs out.

In an another example of SQLI, following test scenario has been used.

- Open Web application
- Start to browse the Web site and go to login page
- Enter Username
- Enter Password: or 1 = 1
- Check whether login has been bypassed or not.

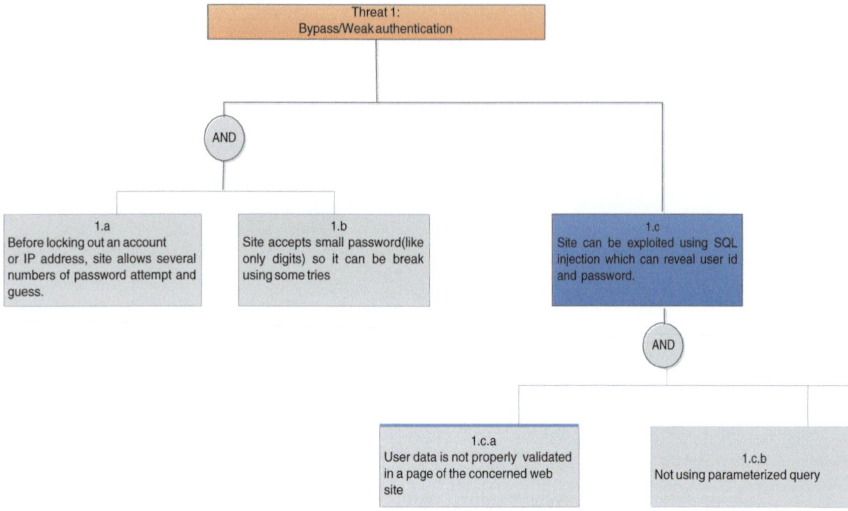

Fig. 7 Threat model diagram of bypass authentication. Adapted from work of Gencer Erdogan [69]

Similarly, several scenarios have been created to test for malicious activities. Security requirements can also be formalized to test the requirements automatically. However, we have not formalized security requirements in this case, but several studies have used this technique [67, 68].

- **Security Testing of Design**: Testing in this phase refers to exposing security vulnerabilities that may enter into application because of design errors such as improper error handling and weak authentication. We have modeled security threats through threat modeling diagram that help experts to discover threats that might enter into the system. We have included a threat model of bypass/weak authentication threat adopted from work of Gencer Erdogan [69] as depicted in Fig. 7. This helps experts to identify whether online shopping application suffers from these threats or not. Similar to this threat model, several component-based threat models have been created to test for any vulnerabilities.
- **Security Testing of Coding**: Security testing in this phase is done to expose any missing secure coding practices or common vulnerabilities pattern. Security experts walked through the whole code to get to know any security flaw or missing secure practices. We have applied static analysis tools such as find-security-bugs [70] to detect any known vulnerabilities. This tool can be added as a plug-in in Findbugs, and it can be used to test for XSS, SQLi, DoS and others in a given program. It is very effective in finding potential XSS such as given in below code 2.

```
<%
String taintedInput = (String)request.getAttribute("input");
%>
 [...]
<%= taintedInput %>
```

In the above code, a potential XSS vulnerability has been found. This vulnerable code could be exploited to execute malicious JavaScript in a client's browser. A list of static analysis tools can be found at site of OWASP [71].

- **Security Testing in Testing Phase**: After review and testing of secure coding, next step is to apply any dynamic security testing tool to identify vulnerabilities during running state of application. These tools find potential vulnerabilities in applications such as input/output validation, specific application problems and others. Dynamic security testing tool such IBM Security AppScan [72] can be applied to uncover security issues. We have downloaded the trail version of this tool from IBM site and tested a dummy Web site "http://www.altoromutual.com" for the security issues. This tool has created several test cases, and some of the test cases are able to exploit the Web site through XSS, SQLi and other vulnerabilities.
  Similar tools based on other security testing techniques such as search-based security testing and concolic testing have been applied in the study. A list of such tools can be found at site of OWASP [73].
- **Security Testing of Deployment and Maintenance**: After deployment of application, health checks of the application and its platform need to be perform periodically to ensure that no new security threats have been introduced in the system. If a system needs any updates or new patches, these should also be properly tested for security vulnerabilities. Experts have proposed several run-time monitoring techniques to monitor the applications at running state. A survey of runtime monitoring techniques can be found in the work of Shahriar and Zulkernine [74].

We have taken a live case of Virtual Freer v1.57 Web application in which very recently SQLi vulnerability has been found. According to vulnerability laboratory [75], an auth bypass session vulnerability has been discovered in the official Virtual Freer v1.57 content management system (CMS). Remote attackers may exploit this vulnerability to access administrator panel or Web user interface. The vulnerability found in a login.php file allows remote attackers to use a 1 = 1 sql payload which leads to bypass validation script at the login.php page. Thus results in access to the administrator panel without a valid account.

If developers have followed the PEST approach as discussed above, then at very early stage they might have tested this vulnerability. We have discussed how a test scenario for the SQLi vulnerability can be created. In case if it has missed in requirement testing, then at design phase this would have been tested using threat model that we have shown in security testing of design.

So the idea behind PEST is to find and fix security issues at the early stages. It seems that these activities will impose overheads in terms of cost, but such activities will prove to be an advantage or value assets in the long run to enable the company having a cutting edge in the competitive market.

8 A Discussion on Industry Practices

Now a days, software is being used in almost every domain, be it entertainment, hospitality, finance and insurance, retail or others. Attackers are targeting their IT environment such as e-commerce, point of sale (POS) and corporate/internal network. Recent study found that in year 2014, finance and insurance industries faced 57 % of their breaches in corporate/internal network and 43 % in e-commerce [17]. This clearly indicates that applications should be properly protected and tested for security issues. As far as security of applications is concerned, Web and mobile applications are prone to severe security flaws. According to report of Trustwave global security [17], Dynamic Application Security Testing (DAST) of Web applications under test revealed 17,748 vulnerabilities, and 98 % of these had one or more security vulnerabilities. Among these applications, 35 % were having information leakage vulnerability, 20 % had XSS vulnerability beside others. This study indicates that security incidents are increasing day by day, and among all exposed asserts, applications are most preferred target for attackers.

In the same study, Trustwave has presented the result of security testing of mobile applications conducted in 2014. They have found at least one vulnerability in 95 % of mobile applications and 6.5 median number of vulnerabilities per application. They have observed 26 % increase over 2013 in critical vulnerability. As mobile applications are continuing to grow, attackers are concentrating more on mobile applications. Inadequate security measures during development will lead to severe setback for the users. So researchers are more concerned about proper security testing of mobile applications.

Generally, security vulnerabilities are tested using application security testing (AST) products. AST includes multiple approaches [76] such as static AST (SAST), dynamic AST (DAST), interactive AST (IAST) that combines SAST and DAST, and mobile AST that combines traditional SAST and DAST along with behavioral analysis. Vendors offer these approaches as a tool or through subscription service. According to study of Gartner [76], majority of enterprises are adopting AST, but they differ on adaptation and maturity of approaches, DAST followed by SAST is the most preferred combination, while IAST and mobile AST are recently emerged.

9 Industry Requirement and Future Trend

With the growing need of Web-based applications and mobile applications, industries are more concerned about protection mechanism. As security incidents are growing day by day, industries are looking forward to develop a comprehensive application security testing platform. There is a need to extend the functionality of security detection with additional functionality, namely security protection of application. Industries are currently practicing on demand-based SaaS:Security as a Service for SAST and DAST, and they are looking forward to support enterprise class requirements with role-based access control (RBAC) to consolidate risk-based reporting [76]. Industries are giving more emphasis on mobile application security since mobile applications are revolutionizing the way industries are doing business, and very quickly, they are connecting to wide customer. Ensuring security in mobile applications is challenge because industries are adopting multiple languages and platforms for their development.

On the basis of recent survey, following future trend is observed:

- Experts are looking on hybrid techniques to construct an automated exploitation toolkit. One such hybrid technique could be symbolic execution and code execution graph analysis with artificial intelligence (AI) techniques [52].
- Researchers are planning to combine protocol state machine with model-based testing and also trying to introduce reverse engineering techniques for automatically extracting the description of input format [77].
- Researchers are working on an emerging technology namely runtime application self-protection (RASP) technology that offers an extension of runtime threat detection and protection. It monitors the execution from inside of the application, controls the application when needed and finally initiates the protection measures [76].
- The major vendors in this market of security testing are IBM, HP, Accenture, Cisco, McAfee, Applause, Veracode and WhiteHat Security. The global security testing market is expected to be doubled by 2019, with an estimated compound annual growth rate (CAGR) of 14.9, and North America is expected to be the largest market for security testing services [78].

10 Conclusion

The rapid increase in the adoption of high functionality software-based applications, complexity and security threats is a reality now. Such a trend, in turn, is bound to throw new areas of exploration with regard to testing from multiple perspectives. Market appears to be shaping a different and inevitable role in the form of security testing services but certainly not without pertinent issues and challenges. A clearly specified understanding of issues and challenges is liable take us in the right direction and to the target effectively and efficiently. Moreover, the present state of the art in security testing appears to be falling short of putting check on the root cause, i.e., inherent vulnerabilities and hence assuring security. Thereby, a framework for phase-based embedded security testing is proposed as continued activity rather than a piecemeal and after-thought approach.

This work is useful for security practitioners to test their application for security vulnerabilities and apply protection accordingly. This will also provide a future direction for security testing researchers as techniques discussed here are good topic to be worked upon specially search-based security testing and hybrid of different techniques.

In fact security threats surrounding the financial applications have increased exponentially and dramatically they continue to evolve. Hence, it is now essentially needed to make software applications highly secure and reliable. Security testing is an effort to reveal those vulnerabilities that may violate state integrity, input validity and logic correctness along with the angle of attack vectors, which may exploit these vulnerabilities. This minimizes risks of security breach and ensures confidentiality, integrity and availability of customer transactions. Modern applications are facing several security challenge prominently include software complexity, third-party code and dynamic security policies. These challenges often exhibit significant impact on applications security.

On the other hand, software is becoming highly complex, huge in size and virtually expandable operational environments. Thus, it is harder to categorically understand applications, difficult to diagnose the problems and test leading to the possibilities of bugs in the system that may likely introduce security vulnerabilities. Because of the market pressure and strict deadlines to produce and deliver software quickly and at a lower cost, third-party code is widely used in software industry. It is a well recognized and obvious fact that third-party code creates many security related issues and concerns that may lead to a serious security risk for companies. Therefore, if third-party code is being used, it must be properly and effectively tested for security flaws. An attempt is made in this chapter to evolve such a mechanism in order to test and assure high level of security at each phase through PEST.

Information security policies specify a set of policies adopted by an organization to protect its information agreed upon with its stakeholder and management. It includes scope of the policy, information classification, management goal of securely handling information at each class and some other. Modern systems interact with external environment where security policies cannot be known well in advance and applied. Thus, security policies are dynamic in nature for such cases like for instance deciding authority of users depending on type of message received through external environment. Therefore, a mechanism is essentially needed that can allow security critical decisions at runtime based on dynamic observations of the environment.

Security is one of the prime aspects of software quality due to prevailing circumstances in software industry. Software could be functionally correct yet, lacking quality in terms of stability, security or its usability. The technology is evolving at incredibly faster pace and spectacularly affecting almost every domain. Therefore, it is indispensable to realize the need and significance to identify a mechanism with comprehensive testing capabilities that is adaptable with evolving dynamic and heterogeneous nature of Web domain. Increasing rate of security breaches has made security testing as a vital part of software life. Software security testing would be instrumental for exposing possible vulnerabilities and associated threats. It provides the assurance that application fulfills current need of security when exposed to a malicious input. All the theories are useless until not cater better tools and techniques. The practices should start right from the beginning at requirement level and continue till the application is finally developed. It would find out vulnerabilities just at the time when they are introduced.

Security testing automation tools need continuous updates with ever evolving and changing technologies. This is one of the major challenges and tools that require adequate integration under existing development workflows. Continuous monitoring and assessment will surely safeguard the upcoming possible threats. The chapter argues and concludes with the fact that security testing is dependent on technology. Hence, it is now necessary to explore testing techniques that can keep track of all possible issues while developing meticulous test design and strategies. Future direction demands how we adapt with secure dynamic pervasive nature of Web.

In view of the current scenario and projections, a bright future appears ahead for quality assurance and software testing domain such as automated, symbolic, formal, phase-based security testing rather than demonstration-based conventional testing. This in turn will be leading to dependable software components in their true spirit to fulfill the future software requirements and secure functionalities. Software testers need to be prepared and be ready to grab the emerging opportunities in the multibillion dollar software security testing industry.

References

1. IDG, Idg study: why application security is a business imperative. White paper, IDG Research Services, 2014
2. Testing guide introduction. https://www.owasp.org/index.php/Testing_Guide_Introduction#Deriving_Functional_and_Non_Functional_Test_Requirements. Accessed 23 Oct 2015
3. B. Potter, G. McGraw, Software security testing. Secur. Priv. IEEE **2**(5), 81–85 (2004)
4. G. Dave, S. Keri, J. Jon, Driving quality, security and compliance in third-party code. http://softwareintegrity.coverity.com/register-for-the-coverity-and-blackduck-webinar.html. Accessed 30 July 2015
5. A. Girard, C. Rommel, Software quality and security challenges growing from rapid rise of third-party code. White paper, VDC Research, 2015
6. H. Janicke, L. Finch, The role of dynamic security policy in military scenarios. in *Proceedings of the 6th European Conference on Information Warfare and Security*, pp. 2007–2009 (2007)
7. I. Magazine and Accenture, Managing complexity still top security challenge, prompting increase in security spend, according to security survey. https://newsroom.accenture.com/news/managing-complexity-still-top-security-challenge//-prompting-increase-in-security-spend-according-to-security-survey.htm. Accessed 30 July 2015

8. B. Schneier, *Beyond fear: thinking sensibly about security in an uncertain world* (Springer Science & Business Media, 2003)
9. D.S. Board, Complex systems can have backdoors and trojan code implanted that is more difficult to find because of complexity. Technical report, Department of Defence, Sept 2007
10. M. Software, More complex = less secure: miss a test path and you could get hacked. Technical report, McCabe Software, Apr 2012
11. F. Brooks, No silver bullet. Apr 1987
12. G. Booch, Object oriented analysis & design with application (Pearson Education India, 2006)
13. R. Bhoraskar, S. Han, J. Jeon, T. Azim, S. Chen, J. Jung, S. Nath, R. Wang, D. Wetherall, Brahmastra: driving apps to test the security of third-party components. in *23rd USENIX Security Symposium (USENIX Security 14)* (USENIX Association, 2014), pp. 1021–1036
14. J. Kirk, Third-party code putting companies at risk. http://www.infoworld.com/article/2626167/data-security/third-party-code-putting-companies-at-risk.html. Accessed 12 Aug 2015
15. Coverity, Study reveals less than fifty percent of third party code is tested for quality and security in development. http://www.coverity.com/press-releases/report-reveals-less-than-fifty-percent-of-third-party//-code-is-tested-for-quality-and-security-in-development. Accessed 12 Aug 2015
16. J. Bayuk, How to write an information security policy. http://www.csoonline.com/article/2124114/strategic-planning-erm/how-to-write-an-information-security-policy.html. Accessed 12 Aug 2015
17. Trustwave, Trustwave global security report. Report, Trustwave, 2015
18. A. Girard, C. Rommel, Application security in the software development lifecycle: issues, challenges and solutions. Report, Quotium, 2015
19. A.R. Nazarov, Protecting your brand. https://searchsecurity.techtarget.com/magazineContent/Protecting-Your-Brand. Accessed 21 Oct 2015
20. S. McConnell, *Code complete*. Microsoft Press (2004)
21. IXIA, Security testing for financial institutions. White paper, IXIA, Jan 2014
22. E.V. Buskirk, Facebook confirms denial-of-service attack (updated) (2009)
23. R. Martin, Your personal data will be stolen (2015)
24. M. Commissioned Forrester Consulting. State of application security. Report, Forrester, Jan 2011
25. M. Howard, S. Lipner, The security development lifecycle: Sdl-a process for developing demonstrably more secure software (2006)
26. G. McGraw, Software security touch point: architectural risk analysis. Technical report, Technical report, 2010. http://www.cigital.com/presentations/ARA10.pdf (2009)
27. OWASP, Comprehensive, lightweight application security process. http://www.owasp.org (2006)
28. C.B. Haley, Arguing security: a framework for analyzing security requirements. PhD thesis, The Open University, 2007
29. G. Sindre, A.L. Opdahl, Capturing security requirements through misuse cases. *NIK 2001, Norsk Informatikkonferanse 2001*. http://www.nik.no/2001 (2001)
30. B. Fabian, S. Gürses, M. Heisel, T. Santen, H. Schmidt, A comparison of security requirements engineering methods. Requirements Eng. **15**(1), 7–40 (2010)
31. G. McGraw, Software security. Secur. Priv. IEEE **2**(2), 80–83 (2004)
32. S. Forum, Architecture and design considerations for secure software, 22 Feb 2011
33. J. Jürjens, Model-based security with umlsec. in *UML Forum, Tokyo* (2003)
34. C. Wysopal, L. Nelson, E. Dustin, D. Dai Zovi. *The Art of Software Security Testing: Identifying Software Security Flaws* (Pearson Education, 2006)
35. SEI CERT coding standards. https://www.securecoding.cert.org/confluence/display/seccode/SEI+CERT+Coding+Standards. Accessed 12 Aug 2015
36. Open source vulnerability database. http://www.osvdb.org/. Accessed 13 Aug 2015
37. K. Spett, Cross-site scripting. SPI Labs **1**, 1–20 (2005)
38. Twitter users including sarah brown hit by malicious hacker attack. http://www.theguardian.com/technology/blog/2010/sep/21/twitter-bug-malicious-exploit-xss. Accessed 13 Aug 2015

39. Sql injection. https://www.owasp.org/index.php/SQL_Injection. Accessed 29 Oct 2015
40. W. Zeller, E.W. Felten, Cross-site request forgeries: exploitation and prevention, Princeton (2008)
41. B. Zeller, Popular websites vulnerable to cross-site request forgery attacks. https://freedom-to-tinker.com/blog/wzeller/popular-websites-vulnerable-cross-site-request-forgery-attacks. Accessed 29 Aug 2015
42. V.D. Gligor, Guaranteeing access in spite of distributed service-flooding attacks. in *Security Protocols* (Springer, 2005), pp. 80–96
43. S. Chen, Application denial of service: is it really that easy? (Hacktics Ltd, 2007)
44. Buffer overflow. https://www.owasp.org/index.php/Buffer_Overflow. Accessed 13 Aug 2015
45. Testing for local file inclusion. https://www.owasp.org/index.php/Testing_for_Local_File_Inclusion.Accessed 29 Oct 2015
46. Imperva, Remote and local file inclusion vulnerabilities 101: and the hackers who love them. Technical report, Imperva, 2012
47. D. Hovemeyer, W. Pugh, Finding bugs is easy. ACM Sigplan Notices **39**(12), 92–106 (2004)
48. C.S. Psreanu, P.C. Mehlitz, D.H. Bushnell, K. Gundy-Burlet, M. Lowry, S. Person, M. Pape, Combining unit-level symbolic execution and system-level concrete execution for testing nasa software. in *Proceedings of the 2008 International Symposium on Software Testing and Analysis* (ACM, 2008), pp. 15–26
49. S. Merz, Model checking: a tutorial overview. in *Modeling and verification of parallel processes* (Springer, 2001), pp. 3–38
50. Static analysis tool. http://babelfish.arc.nasa.gov/trac/jpf/wiki/intro/what_is_jpf. Accessed 13 Aug 2015
51. Barehttp java web server. https://www.savarese.org/software/bare/. Accessed 12 May 2015
52. R. McNally, K. Yiu, D. Grove, D. Gerhardy, *Fuzzing: The State of the Art* (2012)
53. Automated penetration testing with white-box fuzzing. http://msdn.microsoft.com/en-us/library/cc162782.aspx#Fuzzing_topic15. Accessed 17 Dec 2013
54. I. Ciupa, A. Leitner, M. Oriol, B. Meyer, Artoo. in *ACM/IEEE 30th International Conference on Software Engineering, 2008. ICSE'08* (IEEE, 2008), pp. 71–80
55. C. Csallner, Y. Smaragdakis, Jcrasher: an automatic robustness tester for java. Softw. Pract. Exp. **34**(11), 1025–1050 (2004)
56. C. Csalner, Y. Smaragdakis, Check'n'crash: combining static checking and testing. in *Proceedings of the 27th International Conference on Software Engineering* (ACM, 2005), pp. 422–431
57. H. Jaygarl, C.K. Chang, S. Kim, Practical extensions of a randomized testing tool. in *33rd Annual IEEE International on Computer Software and Applications Conference, 2009 (COMPSAC'09)*, vol. 1 (IEEE, 2009), pp. 148–153
58. C. Pacheco, M.D. Ernst, Randoop: feedback-directed random testing for java. in *Companion to the 22nd ACM SIGPLAN Conference on Object-Oriented Programming Systems and Applications Companion* (ACM, 2007), pp. 815–816
59. C.S. Păsăreanu, N. Rungta, W. Visser, Symbolic execution with mixed concrete-symbolic solving. in *Proceedings of the 2011 International Symposium on Software Testing and Analysis* (ACM, 2011), pp. 34–44
60. K. Sen, G. Agha, Cute and jcute: Concolic unit testing and explicit path model-checking tools. in *Computer Aided Verification* (Springer, 2006), pp. 419–423
61. A. Avancini, M. Ceccato, Towards security testing with taint analysis and genetic algorithms. in *Proceedings of the 2010 ICSE Workshop on Software Engineering for Secure Systems* (ACM, 2010), pp. 65–71
62. A. Avancini, M. Ceccato, Comparison and integration of genetic algorithms and dynamic symbolic execution for security testing of cross-site scripting vulnerabilities. Inf. Softw. Technol. **55**(12), 2209–2222 (2013)
63. N. Bhattacharya, A. Sakti, G. Antoniol, Y.-G. Guéhéneuc, G. Pesant, Divide-by-zero exception raising via branch coverage. in *Search Based Software Engineering* (Springer, 2011), pp. 204–218

64. D. Romano, M. Di Penta, G. Antoniol, An approach for search based testing of null pointer exceptions. in *2011 IEEE Fourth International Conference on Software Testing, Verification and Validation (ICST)* (IEEE, 2011), pp. 160–169
65. National vulnerability database. https://web.nvd.nist.gov. Accessed 13 Aug 2015
66. Divide by zero exception ibnstance. http://osvdb.org/show/osvdb/118073. Accessed 13 Aug 2015
67. C.B. Haley, J.D. Moffett, R. Laney, B. Nuseibeh, Arguing security: validating security requirements using structured argumentation (2005)
68. Z. Hui, S. Huang, B. Hu, Y. Yao, Software security testing based on typical ssd: a case study. in *2010 3rd International Conference on Advanced Computer Theory and Engineering (ICACTE),* vol. 2 (IEEE, 2010), pp. V2–312
69. G. Erdogan, Security testing of web based applications (2009)
70. Find security bugs. http://h3xstream.github.io/find-sec-bugs/ Accessed 11 Sep 2015
71. Source code analysis tools. https://www.owasp.org/index.php/Source_Code_Analysis_Tools. Accessed 11 Sept 2015
72. Application security. http://www-03.ibm.com/software/products/en/category/application-security. Accessed 18 Nov 2015
73. Source code analysis tools. https://www.owasp.org/index.php/Category:Vulnerability_Scanning_Tools. Accessed 11 Sept 2015
74. H. Shahriar, M. Zulkernine, Taxonomy and classification of automatic monitoring of program security vulnerability exploitations. J. Syst. Softw. **84**(2), 250–269 (2011)
75. Sql injection vulnerability. http://www.vulnerability-lab.com/get_content.php?id=1592. Accessed 11 Sept 2015
76. J.F. Neil MacDonald, Magic quadrant for application security testing. Technical report, Gartner, 2015
77. F. Pan, Y. Hou, Z. Hong, L. Wu, H. Lai, Efficient model-based fuzz testing using higher-order attribute grammars. J. Softw. **8**(3), 645–651 (2013)
78. Markets and Markets, Security testing market worth $4.96 billion by 2019 (2015)

Uncertainty in Software Testing

Salman Abdul Moiz

Abstract The primary objective of software development is to deliver high quality product at low cost. Testing is inherent in each phase of development as the deliverables of each phase is to be tested to produce a better quality artifact before proceeding to the next phase of development. Software testing describes the discrepancies between the software deliverables and the customer expectations. Software testing life cycle covers test selection, test classification, test execution, and quality estimation. The quality of the deliverable produced may not always be as per the expected outcome or within a probabilistic range. The outcome of testing may be error prone and uncertain because of inadequate techniques for estimation, selection, classification, and execution of test cases. Hence, there is a requirement to model uncertainties after completion of each phase of development. Mechanisms are needed to address uncertainty in each of the deliverables produced during software development process. The uncertainty metrics can help in assessing the degree of uncertainty. Effective modeling techniques for uncertainty are needed at each phase of development.

Keywords Testing · Uncertainty · Modeling uncertainty · Uncertainty principle

1 Introduction

Software development is a complex activity as it is developed for humans by humans. Each phase of software development is dominated by human intervention in making a decision, and hence, the errors are inevitable. The goal of software engineering is to translate requirements into an effective system. In order to achieve quality in software, researchers have developed several techniques to be applied to different types of applications and different scenarios involved in. So, both understanding evolving scenarios during execution of a complex software and

S.A. Moiz (✉)
School of Computer & Information Sciences,
University of Hyderabad, Hyderabad, India
e-mail: salman@uohyd.ac.in

H. Mohanty et al. (eds.), *Trends in Software Testing*,
DOI 10.1007/978-981-10-1415-4_4

selection of appropriate techniques for quality assurance have uncertainty factors attributed with.

Software testing is a continuous mechanism to improve the quality from one phase to other throughout the development of software. Hence, testing is needed for each deliverable produced. The outcome of quality of the deliverables of each phase is often uncertain and unpredictable and at times leads to unavoidable uncertainties. It is observed fact that such uncertainties mostly affect the humans.

Each software engineering activity may be prone to uncertain artifact, and this concept is generally used to conceptualize unpredictability in terms of customer's satisfaction, slippage of schedule, and budget and unpredictable failures in the system. The huge gap between the stakeholders expectations and actual outcomes motivate practitioners to adopt mechanisms to identify and manage software risks and uncertainties.

Today's systems are complex in nature. It is usually a combination of layers, parts, frameworks developed by various organizations. The failure modes of each of the element are not understood, and it is not within ones control. Systems are expected to seamlessly work in presence of varied resources (bandwidth, power, processing power, etc.). The mobility and disconnections characteristics may result in variations of outcomes. The involvement of humans may result in variations in expected outcomes. Hence, there is a need to identify, measure, and manage uncertainty.

Uncertainty modeling approaches are expected to include probabilistic notions of uncertainty. The sources of uncertainty depend on the software artifact being tested. The uncertainty modeling paradigm broadly includes the following steps. Firstly, the sources of uncertainty for the artifact being tested are listed. Secondly, uncertainty is identified when the outcome produced by testing the quality of artifact is neither as per expectation nor is between the expected probabilistic outcomes. The scenarios which lead to uncertainty are analyzed such that proper mechanisms are implemented to reduce uncertain outcomes in future.

The scope of this chapter is to ascertain the uncertainty which exists in various phases of software testing. Since testing is needed at the end of each stage of software development, the causes for such uncertainty varies. With this motive, several sources of uncertainties are presented.

Uncertainty in software engineering is measured using utility curves, anecdotal observation [1], marginal utility [2], and probabilistic assessment [3]. The approaches used to model uncertainty include fuzzification [4], probabilistic reasoning [5], expert systems [6], and neural networks [7]. Each of these approaches addresses uncertainty to specific application, software or activity.

The organization of the chapter is as follows: Section 2 formally defines uncertainty and lists the difference between risk and uncertainty. It also presents various types of uncertainties. The uncertainty principle is stated, and the benefits of uncertainty modeling are presented. In Sect. 3, the sources of uncertainty are described. Section 4 specifies uncertainty in each of the phases of testing activity. Section 5 describes the need for prioritization of uncertainties as the primary concern in software development activity. Section 6 describes the mechanisms available to model some of these uncertainties, and Sect. 7 concludes the chapter by highlighting future scope of work.

2 Uncertainty Preliminaries

The quality of each of the deliverables produced during software developed needs to be assessed. This is possible by testing the observed behavior with expected outcome. To test the quality of each of the artifacts produced, a set of test cases are needed.

A test case is represented as a triplet [I, S, O] where the set of input data given to the system is represented by I, S is the state of the system and O is output or outcome produced by the system (Fig. 1).

State S represents the artifacts of software development produced at the end of each phase. Given finite input set space I and output set space O and a function $f(i)$ which maps a given input to output state, the test case is defined as:

$$\text{Test Case } (\text{I}, \text{S}, \text{O}) : [\forall i \in I, \exists f(i) = r, where\ r \in O]$$

The set of test cases with which the quality of a given deliverable is tested forms test suite. In test oracle, expected outcomes are known. If the output produced by a deliverable or an artifact is the expected one, then the state of the system is certain.

$$\text{Certainty } (\text{I}, \text{S}, \text{O}) : [\exists i \in I \,\&\, o \in O, such\ that\ f(i) = o]$$

Risk is when we are unaware of the outcome, but we do know the distribution of outcomes, i.e., when the expected outcome is not known but the probability of the

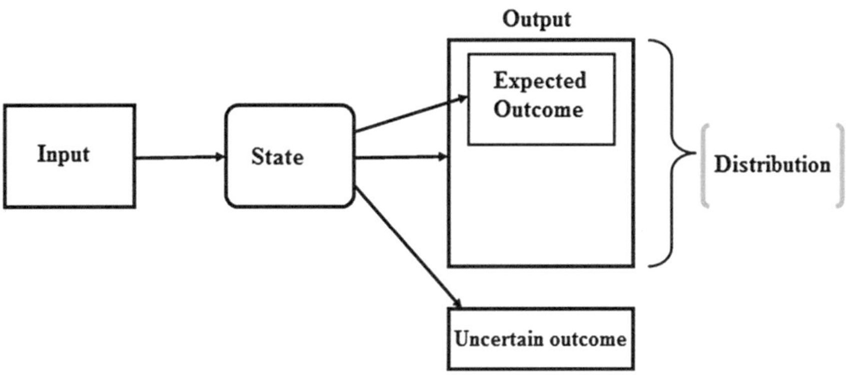

Fig. 1 System under test

outcome is known. For a given input "i," the outcome is unknown but is expected to be within the range or distribution of output space then it leads to risk.

$$\text{Risk }(\mathrm{I, S, O}) : [\exists i \in I, \textit{such that } f(i) = r, \textit{where } r \in O]$$

Uncertainty prevails when neither the outcome nor the distribution is known. The probability of undesired incident is known or justifiable in case of risk, whereas in uncertainty the probability of any such event cannot be computed. For a given input, the outcome produced is outside the output space and then it leads to uncertainty. Uncertainty also exists when the outcome produced varies for the same given input.

$$\text{Uncertainty }(\mathrm{I, S, O}) : [\exists i \in I, \textit{such that } f(i) = p, \textit{where } p \notin O]$$

Robinson et al. [8] formalizes the notion of uncertainty using homomorphism between two objects A and B. The set A represents reality, and the set B is the model developed. If the outputs from both are not the same, then there is error in understanding the application or error in the system developed.

Uncertainty in engineering a system may creep in during requirements engineering and/or system design and development. The study of uncertainty in software development assumes an importance in search of quality software.

2.1 Types of Uncertainty

There are two types of uncertainty as proposed by Luqi and Cooke [9]. One of the uncertainties is to verify whether the given description is the actual specification of the software to be produced. This forms the basis of requirement validation. Huge amount of code generated without knowing the correctness of specifications leads to time and cost overruns.

The second type is related to lifetime of the valid specifications. According to Lehman [10], there are two categories of programs viz., those which are based on given specification and are valid for ever and the one whose specification changes over a period of time.

According to IEEE software engineering standard [11], there are two sources of uncertainty. The first type of uncertainty arises due to the poor understanding of problem domain. The domain experts are needed for the better understanding of problem domain as every problem domain cannot be understood by every person. The second type of uncertainty deals with insufficient emphasis toward maintaining the quality of software and to understand its user needs. The end user expects a quality product satisfying the requirements. Non-conformance of quality of artifacts at each stage of software development may lead to uncertainty.

2.2 *Uncertainty Principle*

Software development is the process of transformation of stakeholders request to a product. Each phase of the development inherently has to deal with ambiguity, incompleteness, or un-deterministic behaviors. Uncertainty is intrinsic in each phase of the development, and the development process has to continue in presence of the uncertainties.

Hardar [5, 12] states the "Maxim of Uncertainty in Software Engineering (MUSE)" as "uncertainty is intrinsic and anticipated in software development activities." This is also referred as "uncertainty principle of software engineering (UPSE)."

The benefits of modeling uncertainty are as follows:

- Uncertainty is a part of software development. If the models can consider representation of software uncertainties, then it would be more accurate. Effective validation of software models also helps in reducing uncertainty in subsequent phases of development.
- The software risk management strategy is effective if the search of uncertainty is carried out accurately. According to Gemmer [13], "It is difficult to reduce risk without knowing the elements of uncertainty. Uncertainty is present in almost all decisions we face." One of the activities to minimize risk is to "systematically search for uncertainty."
- The application of Software uncertainty modeling can improve the software process decisions. Based on the changes in uncertainty levels, the next step of action may be decided in each of the phases.

3 Sources of Uncertainty

Testing becomes a victim of uncertainty. Uncertainty to inputs of a software development phase produces uncertain results. This uncertain results further being used by subsequent phases generate more and more uncertainties to system behavior. Uncertainty creeps in at different phases in software development and may badly affect the software quality. Testing process needs to be aware of such uncertainties involved in software development and corrective measures are to be taken accordingly with an aim to improve the quality of software product. With respect to this objective, it is expected to look for sources that contribute toward uncertainty in software development. Some of the sources include problem domain, software requirements and architecture, solution domain, human participation, requirements churning, learning, cyber physical systems, mobility, and rapid evolution.

The common source of uncertainty in problem domain [12] is from software requirements. Software system models "real-world" problems. As uncertainty prevails in the real world, a system that models such problems also inherits the uncertainties and thereby results in domain uncertainties. When the assumptions, constraints emanating from the requirements of the system are not reflected consistently in the model, it may lead to risks and uncertain outcomes. The models are used to build the system, but as the clarity in understanding requirements can be chaotic or stochastic, it leads to uncertainty in mapping the specifications to models. The models built for real-time application and for safety critical systems suffer from uncertainty characterized by approximation techniques. Models which are built on historical data may not be validated as the structure of model may change over a time due to the varied characteristics and dynamic changes in data set.

Uncertainty exists in stake holder's requests and priorities. Uncertainty arises due to difference in user's view of the requirements and the developers view. There could be alternative architecture solutions for a particular problem. However, if the consequences of various alternatives is not known or cannot be decided, then it leads to uncertainty in freezing architecture of a system [14].

The common paradigms in solution domain are Cyclical and Concurrent Debugging. Traditionally, testing process is cyclic including two stages viz., error finding and error correction. The process repeats until a program gets bug free. This type of debugging is often referred as cyclical debugging [15]. Such a process is suitable for testing of parallel programs. As concurrency is inherent in such environments, it results in different traces of execution in every run for the same inputs. Recreating an execution scenario is probabilistic for possibility of several alternative scenarios concurrency provides.

The cyclical debugging approach cannot be applied for concurrent programs as the behavior of the program for the similar inputs varies when the application is re-executed. The characteristics of concurrency increase the complexity of parallel programs. The effort required to debug parallel programs is higher than that for debugging sequential programs. The concurrent or parallel program does not always result in identical outcome when they are executed with same inputs at several instances. If one process attempt to perform a write operation on to a shared memory location and the second process reads the same data from the same shared space, then the outcome of later operation differs from that of former operation. This variation is due to new or old value acquired by the process.

If the undesirable behavior or the unacceptable outcome arises as a result of similar input, then there is a chance that the program may not be able to create the error situation. The unintended behavior is often referred as Heisenberg's uncertainty principle [16] or Probe effect [17, 18]. Probe effect may not exist if there are no synchronization errors. The non-determinism in concurrent programming is because of the race condition. It is difficult to deal with such non-determinism as program has little or no control over it. The resolution of such issues depends on other attributes. For example, resolution of race may depend on network traffic or on CPU's load. The Heisenberg's uncertainty principle states that "the presence of an observer may affect scientific observation such that absolute confidence in observed results is not guaranteed [12]."

Human participation is indispensable in almost all phases of software development. The participation of human beings introduces uncertainty and unpredictability in software development. Software testing poses more challenges as there will be more involvement of humans in each of the artifact of testing. The knowledge and level of experience of a person varies from one another. Hence, the uncertainty is inevitable. The quality of a deliverable may vary from one team to another if standardized processes are not in place thereby leading to uncertainty.

As proposed by Lehman [10], the validated specification may be axiomatic or non-axiomatic. Axiomatic specification is certain when there is a conformance with the customer regarding the acceptability of requirements and the requirements are validated. The non-axiomatic [19] specification may likely to change. The requirements are not validated because of continuous change in requirements. To deal with such uncertainties, there is a need for periodic assessment of validation of non-axiomatic specification for the entire life of a software product.

Some of the systems developed today use machine learning [20] to realize their functionalities. This may include several cognitive approaches to help users to solve certain complex problems [21]. The learning based systems mostly depend on statistical observations and statistics of past leads to an outcome that is probabilistic.

The realization of cyber physical systems requires complex interactions between software and physical elements [20], for example, security systems, medical devices, and energy distribution. These systems can be uncertain as they rarely capture the dynamic observations of physical environment. Incompleteness in observing behavior of composing elements adds to uncertainty in understanding system dynamics.

A mobile application is expected to be accessed from any place and from any device [20]. Hence, the deployment of application varies from a simple desktop machine to a cloud environment. This includes deployment of applications on mobiles, iPods, and many such handheld devices. The resource availability on each of these devices varies, and it is quiet dynamic in nature. Due to the mobility feature, the outcome of an application may be uncertain for certain target environments.

The evolution of new systems and technologies requires that the existing systems to adapt with these changes. This induces uncertainty in terms of future state of system, type of interfaces, architecture, platform, etc. [20].

4 Uncertainties in Testing Process

Software testing is defined as "the search for discrepancies between the outcome produced by software versus what the user expects it to do" [22]. The process of software testing includes planning, execution, evaluation, and quality estimation. The outcome of testing may be uncertain if the test cases selected are not verified for their correctness. Uncertainty is inherent in software testing but is rarely managed. The primary reason of the uncertainty is human nature of dominance in each activity of software testing.

4.1 Test Planning

Testing goes hand in hand to software development. Outcome of each stage of system development is to be tested for its correctness and consistency. Ideally based on correct output of a phase, the next phase in software development takes place. Each phase of a software process results in certain artifacts, and each artifact is to be tested. This leads to development of quality software. But, the ideal case almost does not exist in reality. Due to the uncertainties involved in its inputs and activities, the correctness of corresponding output cannot be ascertained. It can be predicted. Thus, uncertainty creeps into software development. Based on uncertainty in propositions, design and development of a system can be thought of in different ways. Sequencing of design artifacts and execution logic are greatly influenced by involved uncertainty factors. In order to develop quality software, test planning is to be generated keeping in view the role of uncertainty.

As software testing is human intensive, it introduces uncertainty [5, 12]. The uncertainty in test planning affects the effort and schedule of each artifact. This may result in slippage of schedules and increase in cost of development.

4.2 Test Selection and Classification

Testing is inherently uncertain because only exhaustive testing guarantees confidence about the correctness of results. Test selection activity has to choose finite set of test cases from each of the artifacts. The selection of finite subset of test cases may introduce a level of uncertainty because all the defects may not be detected when using a finite subset.

Selection of test cases and checking for the correctness are the error prone and ambiguous activities in software testing. As exhaustive testing is unrealistic, there is a greater need to identify effective test cases such that system may be adequately tested. Validation depends on fitness and the volume of test cases used. Test case repository may contain redundant, ambiguous, vague, and unfit test cases. This may result in overall increase in effort for software testing. Several attempts were made to address effectiveness, selection, classification, and minimization of software test cases [23]. The techniques for test classification and selection etc., are inadequate and thereby testing quality diminishes due to uncertainty in proper selection of test cases.

Fitness is defined as appropriateness to check quality of software. Measuring the fitness of subset of test cases is a challenging task. Test case fitness depends on several parameters. Ambiguity of fitness of test cases and their fitness parameters have created uncertainty in classification of test cases.

4.3 Test Execution

Test execution process refers to execution of a set of source code with respect to given set of inputs. The system under test may include uncertainties because of difference in outcomes of testing a system in a simulated environment with that of live environment. Test results may vary with respect to person, timing, synchronization, and other dynamical issues.

Uncertainties also emanates from the external software and hardware. According to the report of Therac accidents [24], software error is the primary cause of various accidents and it leads to hazards especially for the mission critical systems. The malfunctioning of a device or hardware is caused because of unexpected outcome of the device driver program. Correcting the errors once found could address the safety issues in mission critical systems. If the causes of a flaw or a bug are properly addressed, the future accidents are likely to be prevented. In this scenario, the real cause of the malfunctioning of the devices is unknown, thereby leading to uncertainty.

A better way to manage uncertainty in testing is to develop more sophisticated oracles [25]. In presence of uncertainty it is difficult to find whether a test case is successful or unsuccessful because the outcome may not be judged properly. There will be inherent risk in test oracle if the distribution is relaxed. However, higher relaxation of test oracle may become barrier in locating faults.

There are two parts of test oracle viz., oracle information and oracle procedure [26]. The oracle information states the condition or statement of correct behavior. The oracle procedure is used to verify the test execution with corresponding oracle. The oracles captured from the formal specification may mostly reflect the intended behavior provided the specifications are correct. Effective test oracles can be directly used during regression testing.

Test oracles may be identified directly from the user acceptance criteria or directly from the specifications. It actually identifies whether a system behaves correctly as expected. The correctness of testing depends on the input set which in turn depends on the type of system being tested. In general, there are two kinds of systems sequential and reactive systems [26]. In sequential systems, only the functional requirements are considered whereas in reactive systems functional and quality requirements are judged.

4.4 Error Tracing

The major challenge in testing is the error tracing. When the outcome of testing an artifact is ambiguous, the cause of erroneous outcome has to be traced. The errors might have emanated from the current phase or any of the previous phases. Effective error tracing is the mechanism that requires that the software artifacts be interrelated so that it can be traced. This is also referred as discovery task [27].

Software traceability is a mechanism which maintains the relationship between one software artifacts to the other [28]. Traceability is a complex activity because of the difficulty in relating the large volumes of interrelated data between one of the artifact to the other.

5 Prioritization of Uncertainty

Though uncertainty permeates software development, it is usually not a first priority [20]. However, while engineering software solutions, it has to be considered as a top concern. This requires a paradigm shift. The following scenario describes the implications of not considering risk as a primary concern. "Distributed Transactions" case study is used to specify the implications of uncertainty [29].

5.1 Distributed Transactions Example

When multiple mobile host accesses shared data from a fixed host at the same time, it may lead to concurrent access and synchronization issues. When the data item(s) is locked by one mobile host, the other mobile has to wait till the resources are unlocked. Consistency of data items is to be assured in presence of the inevitable characteristics of mobile environments like disconnections and mobility.

5.2 Correctness to Utility

Correctness cannot be the end goal in uncertain environments; instead, utility is to be considered. Utility incorporates uncertainty as it allows considering expected outcome. The designers need to trade off the multiple characteristics sacrificing one parameter for the other.

In the distributed transactions example, the consistency can be guaranteed by setting a static timer value. If the transaction under execution is not completed within the stipulated time period, it may be aborted. This helps in solving the problem of starvation. However, this may result in restarting of transaction restarts and wastage of uplink bandwidths. In this scenario, the uplink bandwidth wastage is compromised for starvation issue. The utility is the execution of transaction without starving the resources.

5.3 *Open Loop to Closed Loop*

Open loop systems are traditional, i.e., once software does not give the appropriate solution they are replaced. In closed loop, a system responds dynamically to unexpected outcomes and changes.

In the distributed transactions example, the frequent transaction re-starts may affect the throughput of the system. A mechanism is adopted to keep the uncommitted transactions in queue such that it can be executed later. When the transaction has not completed its execution within the specified time period, the timer value is updated to new value dynamically by increasing it with a step factor "ə." This will enable transaction to be completed when it is scheduled again. The system is closed loop because the timer value is dynamically adjusted based on decision about state of transaction execution.

5.4 *Precise to Approximate*

If an expected behavior of a system is not guaranteed, then there should be mechanisms to ensure that the system works "satisfactorily." This leads to design of approximate solutions.

In the distributed transactions example, time for execution is predicted using soft computing approaches so that decision about execution of a transaction can be taken. There is a possibility that the transaction can be completed within the time period derived from statistical techniques. If the required time is slightly greater than the timer, the transaction still continues execution. In this scenario, though, the expected behavior is not guaranteed, but the transaction execution was acceptable. A decision on allowing transaction execution depends on computing environment at hand during execution. The uncertainty prioritization is to be made based on context as it is shown between estimated time and time taken in reality.

6 Modeling Uncertainties

Uncertainty creeps into the software testing activity because of the failure to measure certain characteristics of software systems and its artifacts. If these characteristics are quantified, then the uncertainty can be addressed in a better way. According to Laplante [30], uncertainty is the inability of quality measure.

Uncertainties that arise out of few characteristics of artifacts under test need to be modeled and managed explicitly. According to Elbaum [25], the three requirements of uncertainty handling in testing are (a) test frameworks that provides automated support for generation of inputs from those distributions that helps in identification of uncertainty, (b) oracles that are used to compare the expected behavior with computed behavior which are probabilistic, and (c) models to represent uncertainties of artifacts under test which also helps in implicit uncertainty quantification.

Traditional software systems run an open loop [20]. In this, the applications are created and deployed and will be used by the stake holders. Once the system does not behave correctly it is replaced with new system. Uncertainty modeling is needed for the open loop systems to save the effort needed to migrate to the new system and/or environment. In closed loop environment, the run time behavior is continuously monitored and the system improves its behavior by dynamically adopting to it. Uncertainty can be handled by using closed loop systems as they allow system to react rapidly to requirement changes, variation in resources, unexpected faults, etc.

Even though it is the known fact that human involvement is the common cause for uncertainly, very few attempts have been made to model them. Risk management has been researched over three decades with signification contributions [31, 32], but baring software dependability and software reliability models [33, 34], uncertainty is often overlooked. The modeling and management of uncertainty is realized using Soft Computing techniques. [35–37].

6.1 Bayesian Approach

Bayesian belief networks are used to model uncertainties in software testing [5, 12]. They are also used in validation of safety critical systems [12]. A Bayesian network is represented as a triplet [V, E, P] in a Directed Acyclic Graph (DAG). V is set of vertices that represent variables (input), E is the set of edges representing causal inferences, and P is the set of probabilities. A Bayesian network represents probability relationship among random variables.

Each edge of a DAG has an associated matrix of probabilities which depicts cause effect relationship, i.e., how value of each variable (cause) affects the probability of each value of other variable (effect). It is basically a parent–child relationship. It is a probabilistic model to know the effect of change in value of a variable on the other.

Each variable V_i of a DAG and takes one value and is assigned with a vector of probabilities labeled Belief (V_i). The probability in Belief (V_i) represents that the variable V_i will take a particular value and directed edge (s_i, d_i) represents causal relationship between source and target node. The causal influence is characterized by conditional probability distribution $P(d_i/s_i)$. The Bayesian network structure is established in consultation with domain experts, and the probabilities of edge matrices can be derived from statistical techniques.

For testing each artifact of software development, a prior belief of the confidence is to be determined. The directed edge from A to B represents that B implements A. The probabilities of correctness are determined by domain experts. The belief vector is revised with repeated computation, and once the correctness of a value is reached, a message is sent to its parent.

The mathematical model provided by Bayesian approach is used for explicitly modeling uncertainties. The graph structure in Bayesian networks corresponds to software systems due to the modularity behavior. The belief values are associated with artifacts, and their relationship is represented by conditional probability matrix. The dynamic changes to software can be captured by Bayesian networks using Bayesian updating.

It is expected that domain experts need to be involved in the entire testing process for establishing belief and the evaluating the associated conditional probability. The challenge is to compute the belief values.

6.2 *Multi-faceted Framework for Test Class Classification*

Effective test case selection and classification helps in proper testing of a system. The objective of the test selection is to select minimum number of test cases such that the effort, costs, and uncertainties are minimized, and at the same time, the effectiveness in testing is achieved [23].

To improve the quality of testing and to minimize the cost and effort of software testing, multi-faceted concepts [38, 39] such as test case fitness evaluation, selection, and classification are to be treated together. Fuzzy logic is a mechanism that helps in solving problems that gives definite outcomes from imprecise, vague, or ambiguous information. It helps in finding a precise solution. Fuzzy logic based multi-faceted measurement framework [23] is used in generating test cases by filtering, prioritizing, and selection. The fuzzy synthesis evaluation algorithm considers multiple criterions to evaluate each test case. The concurrent consideration of fuzzy parameters and testing objectives results in uncertainty.

The multi-faceted measurement is used by tester to generate fitness score of test cases. The grades are assigned to each test case based on the fitness score, and then, the test cases are classified using final grade. The final grades assigned to test cases are on scale of seven [40] worst, worse, bad, average, good, better, excellent.

Gaussian membership function is used to derive the relationship between the grades and fitness parameters by assigning fitness values. Fuzzy feature weighting mechanism is used to address uncertainty.

Multi-objective functions are used to optimize the conflicting test objectives subject to given constraints. The goal is to find acceptable solution by selecting subset of test cases. Test cases are generated by using a multi-objective function by using fuzzy logic and weight assigning method [23]. The fuzzy feature weighting approach considers several parameters which includes fitness value, weight of fitness value, and frequency of parameters in particular module, total number of parameters, and size of test case repository. These parameters vary randomly from one application to another, and the time taken to evaluate the appropriate test cases will be much higher. Further any minor change in the code requires all the parameters to be reassessed. The scale of grade chosen to select test cases is to be standardized.

6.3 Hidden Markov Models

Hidden Markov Models [25] is a probabilistic state machine that can identify veiled properties of system. Elements of HMM include set of states and outcomes represented by "S" and "O," respectively. The total number of states and outcomes is "n." Whenever a state S produces a successful outcome O, the next state produces an observation from O. Transition probability $X = \{x_{i,j}\}$ is assumed to be a constant between all pairs of states i and j. Similarly, emission probability $Y = \{y_{i,j}\}$ represents the possibility of making observation j in state i, which is also believed to be a constant. The states and outcomes of a system can be known but for few applications of HMM, some of the probabilities of X and Y may be unknown. The sequence of outcomes helps in estimating the probability.

The vagueness in inferring the expected outcome is quantified by confusion matrix. Let the system be trained to identify and classify n states and n outcomes. The training of classifiers produces a confusion matrix [25] whose values represent the emission probabilities.

The tester then randomly selects a subset of sample activities such that each succeeding activity is performed with equal probability. Then, based on number of instances of one state producing an outcome, the faults are predicted. The outcomes may also be modeled to specify the degree of correct outcome with a precision value. In Hidden Markov Model, estimation of emission probability is a challenge, and when the states of a system increase, the transition probability changes.

6.4 Rough Sets

Rough sets theory is used in varied applications. It can be used to model and measure uncertainty in applications. Its purpose is to arrive at a proper decision in presence of uncertainty. The approach can be used in many activities of software engineering as for some practical reasons a developer may not be certain of the end product throughout the software development life cycle.

A Conventional set theory for managing incomplete information called Rough set theory was introduced by Pawlak [41]. A rough set represents a function which is used for making decisions based on multiple attributes or with an uncertain information space. Information system development depends on many attributes which can assume values that are uncertain. Hence, use of rough sets in dealing with uncertainty in the development of software systems is promising.

Rough sets can be represented informally in the context of software testing as follows:

Let E be a set specifying expected outcome(s) that an oracle determines for a particular artifact for a given set of input test case I. The conformance of the quality of end product or artifact in concurrence with given set of input(s) can be defined using the following sets:

S1 The outputs produced by testing an artifact are equivalent to the one determined by test oracle for all input test cases

S2 None of the outputs produced by an artifact under test is equivalent to the expected outcomes for all input test cases

S3 The outcomes produced by testing an artifact are possibly equivalent to the output determined by the test oracles

In rough sets, terminology S1 is the lower approximation. The upper approximation is represented as the union of S1 and S3, and the difference between lower and upper approximation represents the boundary region (S4) containing the test cases where the output produced by testing of artifact may be approximately equal to the expected outcome. Since the degree of acceptance of such outcomes cannot be predicted, these test cases are uncertain with respect to the outcomes defined by the test oracles. Rough set is one of the mathematical models to deal with uncertainties [30] by making appropriate decisions.

According to Khoo [42], the data analysis in rough sets theory considers set of objects such that the characteristics of these objects are specified using multi-valued attributes. System behavior can be represented by attributes with multiple values.

Consider a distributed system in which transactions can be initiated by several devices with varied configurations. Conditional attributes specify the data items needed for execution of a particular transaction. The transaction may be completed successfully or may be aborted. Table 1 represents the values of conditional attributes and associated results for transactions. Decision attribute shows either commit or abort of a transactions based on their conditional attributes. The decision attribute "d" assumes value 1 for success and 0 for abort of a transaction.

The confusion is caused due to uncertainty. Confusion or chaos arises when the result of the testing (decision) is different for the same combination of action states.

Based on the commonality of attribute values, we categorize transactions having the same value with respect to an attribute. We call such a set as elementary set. For the attribute $\{a_1\}$, elementary sets are $B_1 = \{r_2, r_6, r_7\}$ and $B_2 = \{r_1, r_3, r_4, r_5\}$. Similarly, for $\{a_2\}$, elementary sets are $B_3 = \{r_3, r_4, r_6\}$ and $B_4 = \{r_1, r_2, r_5, r_7\}$.

Transactions (r_1, r_5) and (r_2, r_7) are identical as they produce the similar outcome for the identical conditional attributes. We categorize such transactions as $A_1 = \{r_1, r_5\}$, $A_2 = \{r_2, r_7\}$, $A_3 = \{r_3\}$, $A_4 = \{r_4\}$, and $A_5 = \{r_6\}$.

Table 1 Transaction status

Transaction	Conditional attributes		Decision attribute (d)
	a1	a2	
r_1	0	0	0
r_2	1	0	0
r_3	0	1	1
r_4	0	1	0
r_5	0	0	0
r_6	1	1	1
r_7	1	0	0

The elementary sets formed by relevant outcome (decision) are called concepts (C) [42]. For the decision attribute 1, $C_1 = \{r_3, r_6\}$ and for the decision attribute 0, $C_2 = \{r_1, r_2, r_4, r_5, r_7\}$. Executions of transactions r_3, r_4 lead to confusion because in r_3, the transaction is committed and in r_4 transaction is aborted for the same conditions. Rough computing is applied to manage these uncertainties.

The lower and upper approximation of the output produced (decision attribute = 1) is evaluated as follows.

Only, $A_5 = \{r_6\}$ is distinguishable from others in C_1 which is r_3. Therefore, lower approximation of C_1 is represented as $\underline{R}(C_1) = \{r_6\}$. Upper approximation is specified as the union of $\underline{R}(C_1)$ and those atoms which are indistinguishable. $\{r_3, r_4\}$ are indistinguishable in C_1. Hence, the upper approximation of C_1 is $\bar{R}(C_1) = \{r_3, r_4, r_6\}$. The boundary region of C_1 is defined as $\bar{R}(C_1) - \underline{R}(C_1)$. Though, the boundary region can be computed manually, but when information space grows, it is difficult to identify the uncertainty in producing an outcome. This scalability issue can be addressed by automating generation of decision rules when the condition attributes are given.

The decision rules are generated using various available mechanisms viz., Exhaustive, LEM2, Genetic and covering algorithms [43]. LEM2 (Learning from Example Module, Version 2) is a rule induction algorithm which uses the idea of multiple attribute value pairs. The requirement information space is given as an input to RSES (Rough Sets Exploration System) tool [44] to generate rules. When the test cases are more, the outcome of the rule generation would be useful in identifying the uncertain outcomes.

Table 1 is given as an input to RSES and LEM2 algorithm is used to generate rules. The rules generated are as follows:

(a) IF (a2 = 0) THEN (d = 0)
(b) IF (a1 = 0) and (a2 = 1) THEN (d = 0)
(c) IF (a1 = 0) and (a2 = 1) THEN (d = 1)
(d) IF (a1 = 1) and (a2 = 1) THEN (d = 1)

The rules (b) and (c) specify the uncertain outcome. If a1 = 0 and a2 = 1, then outcome may be d = 0 or d = 1 which is uncertain. One solution to resolve this uncertainty is inclusion of one of the derived attribute. The uncertainty in the output produced may depend on few characteristics which are not considered in the information system. Table 2 specifies the modification to Table 1 by introducing a derived attribute with an intention to resolve uncertainty.

In the distributed transaction example, the device characteristics (memory capacity, processing speed) are introduced as an additional attribute to make an attempt to address the uncertainty.

By introducing the derived attribute a3, the rules are again generated using LEM2.

(a) IF (a2 = 0) THEN (d = 0)
(b) IF (a2 = 1) and (a3 = 1) THEN (d = 1)
(c) IF (a1 = 0) and (a2 = 1) and (a3 = 0) THEN (d = 0)

Table 2 Test information space with derived attribute

Requirement request	Conditional attributes (action states)			Decision attribute (d)
	a1	a2	a3	
r_1	0	0	1	0
r_2	1	0	1	0
r_3	0	1	1	1
r_4	0	1	0	0
r_5	0	0	1	0
r_6	1	1	1	1
r_7	1	0	1	0

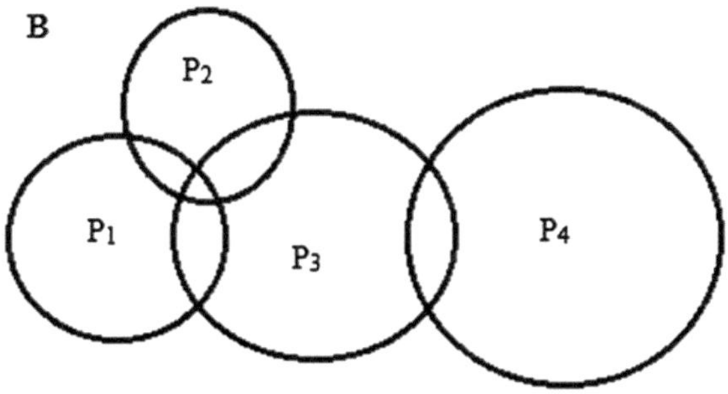

Fig. 2 Quantification of attributes of software processes & tools [47]

Rule (c) specify that if a1 = 0, a2 = 1 and a3 = 0, then the decision parameter is d = 0. By introducing the additional attribute, the uncertainty could be resolved.

The inability to measure few of the characteristics of the software results in uncertainty [30]. While testing some of the real-time applications, the output produced may be different for similar inputs over a period of time. For example, the time for execution of a particular service may be different for the same given inputs. If additional characteristics or relevant parameters are considered, uncertainty can be addressed. Such uncertainties can be estimated using set theoretic approach as described below (Fig. 2).

Let U be the universal set representing quantified properties of software. Let $P_i \subset U$ represents some relevant characteristics of software or its artifact that can be measured and is denoted as P_i. Some of knowledge of all known properties of software or process is $\cup P_i$.

Conversely, $B = U - \cup P_i$ is quantified as the sum of all those properties which are unknown about a process or software which results in uncertainty.

The unknown characteristics of the artifact under test can be computed using B. The subset of such characteristics in addition to the quantified properties considered earlier can help in reducing the uncertainty.

The challenges in managing uncertainty in rough sets with respect to software testing are as follows.

- To predict the derived attribute, related characteristics of parameters need to be considered which may be domain specific and may vary from one artifact to the other. The derived attribute helps in addressing the uncertainty arising due to variation in expected outcome of the quality requirements of a product.
- Similar concept can be extended to all the phases of development provided they are quantified properly.
- The approach to resolve uncertainty is specific to the applications. An attempt can be made to at least have generic guidelines for applications of certain domains.
- The concept of reduct set of rough set theory might help in answering to the questions like "When to stop testing" as there is a possibility for reduction in the information space.

6.5 Design Decisions Under Uncertainty

According to Frank [45], the information extracted from the models must correspond to the information extracted for corresponding operations on reality. Software design decisions are dependent on multiple goals that are difficult to characterize, thereby leading to conflicts. The goals include the non-functional concerns such as security, reliability, and performance of software. Classifying the goals and evaluating the trade of each of the alternatives help in software design decisions. To handle the complexity of the system design decisions, Emmanuel [14] proposed the following process:

The first step is to define a multi-objective architecture decision model consisting of design decisions, dependency constraints, model parameters, and optimization goals. The complexity of the design increases when the number of objectives grows.

The second step is to define the cost benefit model that allows system designers to map design decisions and satisfaction criterion to financial goals that are of utmost importance to its customers. In reality, it is too hard to quantify them.

In the third step, design decision risk is addressed and quantified. Measures related to net benefit can be of interest to an architect. In [14], the goal failure and project failure risks are addressed. Finally, candidate architecture is selected based on comparison of risks of possible other design architectures.

The design decision can be open or closed. A design decision is said to be closed if all the architectures shortlisted agree on the option selected for a particular decision. The design decision is said to be open, if the shortlisted architectures has the provision of selecting alternate options. In case of having multiple objectives playing role in making design decisions, standard methods for multi-criteria decision making are to be applied. However, this requires identifications of right objectives that are relevant to system design decision and more important is their quantifications.

Study on the impact uncertainty makes in system design is essential to evaluate the risk involved in development of such systems.

7 Conclusion

Uncertainty in software development permeates not only in understanding the problem and domain modeling but also exists in design and implementation. As each artifact of software process is to be tested, there is a possibility of uncertainty in each of these phases. Few of the sources of uncertainty are discussed. Testing performed by considering the uncertainty might help to identify more and more such sources. The source of uncertainty varies from one domain to another and one model to other. Uncertainty in test planning, test selection, and test execution is discussed. Few of the probabilistic approaches for modeling uncertainty are presented. Uncertainty metrics play a vital role in calculating the degree of error present in deliverables. Some of the approaches for quantifying uncertainties are presented here.

Future Directions: Software uncertainty is to be considered as a primary concern in development of software systems. This induces a paradigm shift [20]. Each phase of software development has to model uncertainty so that uncertainty is assessed in each phase and measures are to be taken such that it does not get propagate to other phases. This is possible when the deliverables are tested for the expected outcomes. Reducing uncertainty in software development cycle reduces risks involved with deliverables.

The V model [46] is to be refined to check for uncertainty in each phase of testing, i.e., during verification phase and finally at the validation phase.

The uncertainty metrics are needed at the requirement specification, design, and coding phases. The metrics available in literature do not apply to all phases of development. As uncertainty may propagate from one phase to another, there is a need to assess propagation metrics such that the uncertainty emanating from previous phase can be considered for evaluating uncertainty in successive phases. As uncertainty cannot be eliminated totally, the approaches to reduce uncertainty have to be adopted dynamically in course of software development.

There is a need for richer testing frameworks which uses the distributions of inputs such that the distributions can help in identifying the uncertainty. It will be difficult to assess each and every discrete input for identification of uncertainty. Hence, there is a need for devising a strategy that identifies such input distributions. Further, the test oracles have to be stronger such that the acceptable and unacceptable behaviors can be distinguished, and in case of mismatch, it can be quantified.

Testing uncertainty in software development as we see is an important issue. So, this makes the need for development of tools that considers role of uncertainty in software development.

References

1. W. Royce, Measuring agility and architectural integrity. Int. J. Softw. Inform. **5**(3), 415–433 (2011)
2. R.B. Paramkushan, A decision theoretic model for information technology. Manag. Stud. **1**(1), 47–63 (2013)
3. B. Kitchenham, S. Linkman, Estimates, uncertainty and risk. IEEE Softw. **14**(3), 69–74 (1997)
4. H. Madsen, P. Thyregod, B. Burtschy, G. Albenau, F. Popentiu, A fuzzy logic approach to software testing & debugging, in *Safety & Reliability for Managing Risk* (*ESREL 2006*), vol. 11, ed. by C. Guedes Soares, E. Zio (Taylor & Francisc Group, Abingdon, 2006), pp. 1435–1442
5. H. Ziv, D.J. Richardson, Bayesian-network confirmation of software testing uncertainties, in *Proceedings of Sixth European Software Engineering Conference* (*ESEC*), 1997
6. R. Buxton, Modeling uncertainty in expert systems. Int. J. Man Mach. Stud. **31**(4), 415–476 (1989)
7. W.A. Wrigh, G. Ramage, D. Cornford, I.T. Nabney, Neural network modeling with input uncertainty: theory & applications. J. VLSI Signal Process. Syst. Signal Image Video Technol. **26**(1), 169–188 (2000)
8. V.B. Robinson, A.U. Frank, About kinds of uncertainty in collection of spatial data, in *Proceedings of AUTO-CARTO 7* (1985), pp. 440–449
9. Luqi, D. Cooke, The management of uncertainty in software development, in *Proceedings of 16th International Conference on Computer Software & Applications* (1992), pp. 381–386
10. M.M. Lehman, Programs, life cycles and laws of software evolution. Proc. IEEE **68**(9), 1060–1075 (1980)
11. Computer Society IEEE, IEEE guide for the use of IEEE standard dictionary of measures to produce reliable software, in *IEEE Standards Collection Software Engineering* (IEEE Computer Society Press, New York, 1994)
12. H. Ziv, D.J. Richardson, The uncertainty principle in software engineering, in *Proceedings of 19th International Conference on Software Engineering*, 1996
13. A. Gemmer, Risk management: moving beyond process. IEEE Comput. **30**(5), 33–43 (1997)
14. E. Letier, D. Stefan, E.T. Barr, Uncertainty, risk and information value in software requirements & architecture, in *Proceedings of 36th International Conference on Software Engineering* (2014), pp. 883–894
15. C.E. Medowell, D.P. Helmbold, Debugging concurrent programs. ACM Comput. Surv. **21**(4), 593–628 (1989)
16. C.H. Ledoux, D.S. Paker Jr, Saving traces for Ada debugging, in *Proceedings of the Ada International Conference ACM* (1985), pp. 97–108
17. Jason Gait, A probe effect in concurrent programs. Softw. Pract. Exp. **16**(3), 225–233 (1986)
18. J. Gait, A debugger for concurrent programs. Softw. Pract. Exp. **15**(6), 539–594 (1985)
19. C.V. Ramamoorthy, Danilel E. Cooke, The Correspondence between Methods of Artificial Intelligence and the Production and Maintenance of Evolutionary Software, in *Proceedings of the 3rd International IEEE Conference on Tools for Artificial Intelligence* (1991), pp. 114–118
20. D. Garlan, Software engineering in an uncertain world, in *ACM Proceedings of FOSER* (2010), pp. 125–128
21. D. Garlan, B. Schmerl, The RADAR architecture for personal cognitive assistance. Int. J. Softw. Eng. Knowl. Eng. **17**(2) (2007)

22. A.L. Goel, Software reliability models: assumptions, limitations and applicability. IEEE Trans. Softw. Eng. **SE-11**(12),1411–1423 (1985)
23. M. Kumar, A. Sharma, R. Kumar, multi-faceted measurement framework for test case classification & fitness evaluation using fuzzy logic based approach. Chiang Mai J. Sci. **39**(3), 486–497 (2012)
24. N.G. Leveson, C.S. Turner, An investigation of the Therac-25 accidents. IEEE Comput. **26**(7), 18–41 (1993)
25. S. Elbaum, D.S. rosenblum, Known unknowns: testing in the presence of uncertainty, in *Proceedings of FSE* (2014), pp. 833–836
26. D.J. Richardson, S.L. Aha, T.O. O'Malley, Specification based test oracles for reactive systems, in *Proceedings of 14th International Conference on Software Engineering* (*ICSE*) (1992), pp. 105–118
27. P.T. Devanbu, R.J. Brachman, P.J. Selfridge, B.W. Ballard, LaSSIE: a knowledge based software information systems, in *Proceedings of 12th International Conference on Software Engineering* (1990), pp. 249–261
28. A.M. Davis, Tracing: a simple necessity neglected. IEEE Softw. **12**(5), 6–7 (1995)
29. S.A. Moiz, R. Lakshmi, Single lock manager approach for achieving concurrency in mobile environments, in *14th IEEE International Conference on High Performance Computing* (*HiPC*) (Springer LNCS 4873, 2007), pp. 650–660
30. P.A. Laplante, C.J. Neil, Modeling uncertainty in software engineering using rough sets. Innovations Syst. Softw. Eng. **I**, 71–78 (2005)
31. B.W. Boehm, *Software Risk Management* (IEEE Computer Society Press, Washington, D.C., 1989)
32. B.W. Boehm, Software risk management: principles and practices. IEEE Softw. **8**(1), 32–41 (1991)
33. B. Littlewood, How to measure software reliability and how not to. IEEE Trans. Reliab. **R-28** (2), 103–110 (1979)
34. B. Littlewood, L. Strigini, Validation of ultrahigh dependability for software-based systems. Commun. ACM **36**(11), 69–80 (1993)
35. J. Pearl, *Probabilistic reasoning in intelligent systems: networks of plausible inference* (Morgan Kaufmann Publishers, San Mateo, 1988)
36. R.E. Neapolitan, *Probabilistic reasoning in expert systems: theory and algorithms* (Wiley, New York, 1990)
37. D.E. Heckerman, A. Mamdani, M.P. Wellman, Real-world applications of Bayesian networks. Uncertainty AI Commun. ACM **38**(3), 24–26 (1995)
38. M. Kumar, A. Sharma, R. Kumar, Towards multi-faceted test cases optimization. J. Softw. Eng. Appl. **4**(9), 550–557 (2011)
39. M. Kumar, A. Sharma, R. Kumar, Soft computing-based software test cases optimization: a survey. Int. Rev. Comp. Softw. **6**(4), 512–526 (2011)
40. S. Yogesh, A. Kaur, B. Suri, Test case prioritization using ant colony optimization. ACM SIGSOFT Softw. Eng. Notes **35**(4), 1–7 (2010)
41. Z. Pawlak, Rough sets. Int. J. Comput. Inform. Sci. **11**(5), 341–356 (1982)
42. L.P. Khoo, S.B. Tor, L.Y. Zhai, A rough-set-based approach for classification and rule induction. Int. J. Adv. Manuf. Technol. **15**(7), 438–444 (1999)
43. J.W. Grazymala-Busse, A new version of the rule induction system lers. J. Fundamenta Informaticae **31**(1), 27–39 (1997)
44. http://rses.software.informer.com/
45. A. Frank, Conceptual framework for land information system—a first approach, in *Proceedings of Commission 3 of the FIG* (1982)
46. K. Forsberg, H. Mooz, The relationship of system engineering to the project cycle, in *Proceedings of First Annual Symposium of National Council on System Engineering* (1991), pp. 57–65
47. PA Laplante, CJ Neils, Uncertainty: a meta-property of software, in *Proceedings of the IEEE/NASA 29th Software Engineering Workshop* (2005), pp. 228–233

Separation Logic to Meliorate Software Testing and Validation

Abhishek Kr. Singh and Raja Natarajan

Abstract The ideal goal of any program logic is to develop logically correct programs without the need for predominant debugging. Separation logic is considered to be an effective program logic for proving programs that involve pointers. Reasoning with pointers becomes difficult especially due to the way they interfere with the modular style of program development. For instance, often there is aliasing arising due to several pointers to a given cell location. In such situations, any alteration to that cell in one of the program modules may affect the values of many syntactically unrelated expressions in other modules. In this chapter, we try to explore these difficulties through some simple examples and introduce the notion of separating conjunction as a tool to deal with it. We introduce separation logic as an extension of Hoare Logic using a programming language that has four pointer-manipulating commands. These commands perform the usual heap operations such as lookup, update, allocation and de-allocation. The new set of assertions and axioms of separation logic are presented in a semi-formal style. Examples are given to illustrate unique features of the new assertions and axioms. Finally, the chapter concludes with proofs of some real programs using the axioms of separation logic.

Keywords Modular testing · Formal methods · Separation logic · Pointer aliasing · Hoare logic · Programming methodology

A.Kr. Singh (✉) · R. Natarajan
School of Technology & Computer Science, Tata Institute of Fundamental Research, Mumbai 400005, India
e-mail: abhishek.uor@gmail.com
URL: http://www.tcs.tifr.res.in/abhishek

R. Natarajan
e-mail: raja@tifr.res.in
URL: http://www.tcs.tifr.res.in/raja

H. Mohanty et al. (eds.), *Trends in Software Testing*,
DOI 10.1007/978-981-10-1415-4_5

1 Introduction

The ideal goal of any program logic is to help in developing logically correct programs without the need for predominant debugging. *Separation logic* [1] is one such Program Logic. It can be seen as an extension to standard *Hoare Logic* [2]. The aim of this extension is to simplify reasoning with programs that involve low-level memory access, such as *pointers*.

- Reasoning with pointers becomes very difficult because of the way they interfere with the modular style of program development.

Structured programming approaches provide the freedom to develop a large program by splitting it into small modules. Hence, at any given time, a programmer can only concentrate on developing a particular module against its specification. In the absence of pointer commands, the proof of correctness of these small modules can easily be extended to a global proof for the whole program.

- For example, consider the following specification expressed in the form of a *Hoare triple*:

$$\{x = 4\} \quad \texttt{x := 8} \quad \{x = 8\}$$

It claims that if execution of the command `x:= 8` begins in a state where $x = 4$ is true, then it ends in a state where $x = 8$ is true. This seems trivially true. In a similar way, one can easily verify the validity of following specification:

$$\{x = 4 \wedge y = 4\} \quad \texttt{x := 8} \quad \{x = 8 \wedge y = 4\}$$

Here, the proposition $y = 4$ should remain true in the postcondition, since the value of variable y is not affected by the execution of command `x:= 8`.

- The reasoning above is an instance of a more general rule of Hoare logic, viz. the *rule of constancy*

$$\frac{\{p\}\, c\, \{q\}}{\{p \wedge r\}\, c\, \{q \wedge r\}}$$

where no free variable of r is modified by c. It is easy to see how this rule follows from the usual meaning of the assignment operator. The following sequence of state transitions can be used to illustrate the above idea:

Local Reasoning		Global Reasoning
Store: x:4 `x:=8;` ↓ Store: x:8	$\Longrightarrow$	Store: x:4, y:_ `y:= 4;` ↓ Store: x:4, y:4 `x:= 8;` ↓ Store: x:8, y:4

However, if the program modules use data structures such as arrays, linked lists or trees, which involve addressable memory, then extending local reasoning is not so easy using the rule of constancy.

- For example, consider a similar specification which involves mutation of an array element

$$\{a[i] = 4 \wedge a[j] = 4\}\; \texttt{a[i]} := 8\; \{a[i] = 8 \wedge a[j] = 4\}$$

In order to realize that the above is not a valid specification, it is enough to consider the case when $i = j$. Therefore, to avoid this problem, an extra clause $i \neq j$ is needed in the precondition to make it a valid specification. To complicate the situation even further, consider the following specification:

$$\{a[i] = 4 \wedge a[j] = 4 \wedge a[k] = 4\}\; \texttt{a[i]} := 8\; \{a[i] = 8 \wedge a[j] = 4 \wedge a[k] = 4\}$$

In this case, two more clauses, i.e., $i \neq j$ and $i \neq k$ are needed in the precondition to make it a valid specification. These are the extra clauses that programmers often forget to mention. However, these clauses are necessary since it assures that the three propositions $a[i] = 4, a[j] = 4$ and $a[k] = 4$ refer to mutually disjoint portions of heap memory, and hence, mutating one will not affect the others. Thus, although $\{a[i] = 4\}$ `a[i]:= 8`$\{a[i] = 8\}$ is a valid specification, applying rule of constancy in these cases can lead us to an invalid conclusion.

These kinds of non-sharing are often assumed by programmers. However, in classical logic non-sharing needs explicit mention, which results in a program specification that looks clumsy.

- Separation logic deals with this difficulty by introducing a separating conjunction, $P * Q$, which asserts that P and Q hold for disjoint portions of addressable memory.
- In this sense, it is closer to the programmer's way of reasoning.

Since non-sharing is the default in separating conjunction, the above specification can be written succinctly as:

$$\{a[i] \mapsto 4 * a[j] \mapsto 4 * a[k] = 4\}\; \texttt{a[i]} := 8\; \{a[i] \mapsto 8 * a[j] \mapsto 4 * a[k] \mapsto 4\}$$

where $p \mapsto e$ represents a single-cell heaplet with p as domain and e is the value stored at the address p. Thus, the assertion $a[i] \mapsto 4 * a[j] \mapsto 4$ means that $a[i] \mapsto 4$ and $a[j] \mapsto 4$ holds on disjoint parts of the heap and hence $i \neq j$. Although the normal rule of constancy is no more valid, we have the following equivalent rule called "*frame rule*"

$$\frac{\{p\}\,c\,\{q\}}{\{p * r\}\,c\,\{q * r\}}$$

where no variable occurring free in r is modified by c.

In this section, we have seen some specifications and their meanings in an informal way. We need to define these notions formally, before we can discuss new assertions and other features of separation logic more rigorously. Section 2 prepares this background by defining a language $\mathcal{L}$ and introducing Hoare Logic. Section 3 introduces the new forms of assertions in separation logic. It also extends axioms of Hoare Logic to include some new axioms for reasoning with pointers. In Sect. 4, we describe the idea of annotated proofs and present the proof of an in-place list reversal program using the axioms of separation logic.

2 Background

In this section, we specify a language $\mathcal{L}$ by defining its syntax and semantics. A subset S of the language $\mathcal{L}$ is then used to introduce the axioms of Hoare Logic for commands that do not involve pointers. We now describe the structure and meaning of various commands in the language $\mathcal{L}$:

- **Skip**.
  Command: `Skip`
  Meaning: The execution has no effect on the state of computation.
- **Assignment**.
  Command: `x := e`
  Meaning: The command changes the state by assigning the value of term e to the variable x.
- **Sequencing**.
  Command: $C_1; \ldots; C_n$
  Meaning: The commands $C_1, \ldots, C_n$ are executed in that order.
- **Conditional**.
  Command: `if` `b` `then` C_1 `else` C_2
  Meaning: If the boolean expression b evaluates to **true** in the present state, then C_1 is executed. If b evaluates to **false**, then C_2 is executed.
- **While Loop**.
  Command: `while` `b` `do` C
  Meaning: If the boolean expression b evaluates to **false,** then nothing is done. If b evaluates to **true** in the present state, then C is executed and the while command is then repeated. Hence, C is repeatedly executed until the value of b becomes **false**.

- **Allocation**.
  Command: $\texttt{x} := \texttt{cons}(e_1, \ldots, e_n)$
  Meaning: The command $\texttt{x} := \texttt{cons}(e_1, \ldots, e_n)$ reserves n consecutive cells in the memory initialized to the values of $e_1, \ldots, e_n$, and saves in x the address of the first cell. Note that, for successful execution of this command, the addressable memory must have n uninitialized and consecutive cells available.
- **Lookup**.
  Command: $\texttt{x} := [\mathsf{e}]$
  Meaning: It saves the value stored at location e in the variable x. Again, for the successful execution of this command, location e must have been initialized by some previous command of the program. Otherwise, the execution will abort.
- **Mutation**.
  Command: $[\mathsf{e}] := \mathsf{e}'$
  Meaning: The command $[\mathsf{e}] := \mathsf{e}'$ stores the value of expression e' at the location e. Again, for this to happen, location e must be an active cell of the addressable memory.
- **Deallocation**.
  Command: `free (e)`
  Meaning: The instruction `free (e)` deallocates the cell at the address e. If e is not an active cell location, then the execution of this command shall abort.

2.1 Formal Syntax

The structure of commands in the language $\mathcal{L}$ can also be described by the following abstract syntax:

Syntax of Commands

$$
\begin{aligned}
\langle cmd \rangle ::= \; & \text{`}\texttt{skip}\text{'} \\
| \; & \langle var \rangle := \langle aexp \rangle \\
| \; & \langle cmd \rangle \text{ `}\texttt{;}\text{' } \langle cmd \rangle \\
| \; & \text{`}\texttt{if}\text{' } \langle bexp \rangle \text{ `}\texttt{then}\text{' } \langle cmd \rangle \text{ `}\texttt{else}\text{' } \langle cmd \rangle \\
| \; & \text{`}\texttt{while}\text{' } \langle bexp \rangle \text{ `}\texttt{do}\text{' } \langle cmd \rangle \\
| \; & \langle var \rangle := \text{`}\texttt{cons}\text{' } (\langle aexp \rangle, \ldots, \langle aexp \rangle) \\
| \; & \langle var \rangle := [\langle aexp \rangle] \\
| \; & [\langle aexp \rangle] := \langle aexp \rangle \\
| \; & \text{`}\texttt{free}\text{' } (\langle aexp \rangle)
\end{aligned}
$$

where *aexp* and *bexp* stand for arithmetic and boolean expressions, respectively. The syntax of these is as follows:

$$
\begin{aligned}
\langle aexp \rangle ::= &\ \mathbf{int} \\
|\ & \langle var \rangle \\
|\ & \langle aexp \rangle + \langle aexp \rangle \\
|\ & \langle aexp \rangle \text{ - } \langle aexp \rangle \\
|\ & \langle aexp \rangle \times \langle aexp \rangle
\end{aligned}
$$

$$
\begin{aligned}
\langle bexp \rangle ::= &\ \mathbf{true} \mid \mathbf{false} \\
|\ & \langle aexp \rangle = \langle aexp \rangle \\
|\ & \neg \langle bexp \rangle \\
|\ & \langle bexp \rangle \wedge \langle bexp \rangle \\
|\ & \langle bexp \rangle \vee \langle bexp \rangle \\
|\ & \langle bexp \rangle \Rightarrow \langle bexp \rangle
\end{aligned}
$$

2.2 *Formal Semantics*

The formal semantics of a programming language can be specified by assigning meanings to its individual commands. A natural way of assigning meaning to a command is by describing the effect of its execution on the *state* of computation. The state of a computation can be described by its two components, *store* and *heap*.

- Store, which is sometimes called stack, contains the values of local variables. Heap maintains the information about the contents of active cell locations in the memory. More precisely, both of them can be viewed as partial functions of the following form:

$$\mathtt{Heaps} \triangleq \mathtt{Location} \rightharpoonup \mathtt{Int} \quad \mathtt{Stores} \triangleq \mathtt{Variables} \rightharpoonup \mathtt{Int}$$

- Note that the notations, **cons** and [-], which refer to the heap memory, are absent in the syntax of *aexp*. Therefore, the evaluation of an arithmetic or boolean expression depends only on the contents of the store at any given time. We use the notation $s|{=}e \Downarrow v$ to assert that the expression e evaluates to v with respect to the content of store s. For example, let $s = \{(x,2),(y,4),(z,6)\}$ then $s|{=}x \times (y+z) \Downarrow 20$. For our discussion, we assume that this evaluation relation is already defined.

Operational Semantics: We now define a transition relation, represented as $\langle c,(s,h)\rangle \rightarrowtail (s',h')$, between states. It asserts that, if the execution of command c starts in a state (s,h), then it will end in the state (s',h'). The following set of rules, SEMANTICS-I and SEMANTICS-II, describe the operational behavior of every command in the language $\mathcal{L}$, using the transition relation $\rightarrowtail$.

SEMANTICS-I (Commands without pointers)

$$\text{Skip}\ \langle skip, (s,h)\rangle \rightarrowtail (s,h)$$

$$\text{Asign}\frac{s|{=}e \Downarrow v}{\langle \mathtt{x := e}, (s,h)\rangle \rightarrowtail (s[x:v],h)}$$

$$\text{Seq}\frac{\langle c_1, \rangle(s,h)\rangle \rightarrowtail (s',h')}{\langle c_1;c_2,(s,h)\rangle \rightarrowtail \langle c_2,(s',h')\rangle}$$

$$\text{If}-\text{T}\frac{s|{=}e \Downarrow \textbf{true} \quad \langle c_1,(s,h)\rangle \rightarrowtail (s',h')}{\langle \mathtt{if}\, e\, \mathtt{then}\, c_1\, \mathtt{else}\, c_2, (s,h)\rangle \rightarrowtail (s',h')}$$

$$\text{If}-\text{F}\frac{s|{=}e \Downarrow \textbf{false} \quad \langle c_2,(s,h)\rangle \rightarrowtail (s',h')}{\langle \mathtt{if}\, e\, \mathtt{then}\, c_1\, \mathtt{else}\, c_2, (s,h)\rangle \rightarrowtail (s',h')}$$

$$\text{W}-\text{F}\frac{s|{=}e \Downarrow \textbf{false}}{\langle \mathtt{while}\, e\, \mathtt{do}\, c, (s,h)\rangle \rightarrowtail (s,h)}$$

$$\text{W}-\text{T}\frac{s|{=}e \Downarrow \textbf{true} \quad \langle c,(s,h)\rangle \rightarrowtail (s',h') \quad \langle \mathtt{while}\, e\, \mathtt{do}\, c, (s',h')\rangle \rightarrowtail (s'',h'')}{\langle \mathtt{while}\, e\, \mathtt{do}\, c, (s',h')\rangle \rightarrowtail (s'',h'')}$$

SEMANTICS-II (Commands with pointers)

$$\text{Alloc}\frac{s|{=}e_1 \Downarrow v_1,\ldots,s|{=}e_n \Downarrow v_n \quad l,\ldots,l+n-1 \in \mathtt{Locations} - \mathtt{dom}\, h}{\langle \mathtt{x := cons}(e_1,\ldots,e_n),(s,h)\rangle \rightarrowtail (s[x:l],h[l:v_1,\ldots,l+n-1:v_n])}$$

$$\text{Look}\frac{s|{=}e \Downarrow v \quad v \in \mathtt{dom}\, h}{\langle \mathtt{x := [e]}, (s,h)\rangle \rightarrowtail (s[x:h(v)],h)} \qquad \frac{s|{=}e \Downarrow v \quad v \notin \mathtt{dom}\, h}{\langle \mathtt{x} := [e],(s,h)\rangle \rightarrowtail \textbf{abort}}$$

$$\text{Mut}\frac{s|{=}e \Downarrow v \quad v \in \mathtt{dom}\, h \quad s|{=}e' \Downarrow v'}{\langle [\mathtt{e}] := e', (s,h)\rangle \rightarrowtail (s,h[v:v'])} \qquad \frac{s|{=}e \Downarrow v \quad v \notin \mathtt{dom}\, h}{\langle [e] := e',(s,h)\rangle \rightarrowtail \textbf{abort}}$$

$$\text{Free}\frac{s|{=}e \Downarrow v \quad v \in \mathtt{dom}\, h}{\langle \mathtt{free(e)},(s,h)\rangle \rightarrowtail (s,h\rceil(\mathtt{dom}\, h - \{v\}))} \qquad \frac{s|{=}e \Downarrow v \quad v \notin \mathtt{dom}\, h}{\langle \text{free}(e),(s,h)\rangle \rightarrowtail \textbf{abort}}$$

where $f[x:v]$ represents a function that maps x to v and all other arguments y in the domain of f to fy. Notation $f \ominus A$ is used to represent the restriction of function f to the domain A.

- An important feature of the Language $\mathcal{L}$ is that any attempt to refer to an unallocated address causes the program execution to **abort**. For example, consider the following sequence of commands.

		Store x:10, y:0 Heap empty
Allocation	`x:= cons(1,2);`	↓
		Store x:10, y:0 Heap 10:1, 11:2
Lookup	`y:= [x];`	↓
		Store x:10, y:10 Heap 10:1, 11:2
Deallocation	`free(x+1);`	↓
		Store x:10, y:10 Heap 10:1
Mutation	`[x+1]:= y;`	↓
		abort

Here, an attempt to modify the content of address 11 causes the execution to **abort** because this location was deallocated by the previous instruction.

2.3 *Hoare Triples*

The operational semantics of language $\mathcal{L}$ can be used to prove any valid specification of the form $\langle c, (s, h)\rangle \overset{*}{\rightarrowtail} (s', h')$. However, this form of specification is not the most useful one. Usually, we do not wish to specify programs for single states. Instead, we would like to talk about a set of states and how the execution may transform that set. This is possible using a Hoare triple $\{p\}c\{q\}$,

- Informally, it says that if the execution of program c begins in a state that satisfies proposition p, then if the execution terminates it will end in a state that satisfies q.

where p and q are assertions that may evaluate to either **true** or **false** in a given state. We will use notation $[\![p]\!]s\,h$, to represent the value to which p evaluates, in the state (s, h). Therefore,

$$\{p\}c\{q\}\text{holds iff}\forall(s,h) \in \texttt{States}, [[p]]s\,h \Rightarrow$$
$$\neg(\langle c, (s,h)\rangle \overset{*}{\rightarrowtail} \textbf{abort}) \wedge (\forall(s',h') \in \texttt{States}, (\langle c, (s,h)\rangle \overset{*}{\rightarrowtail} (s',h')) \Rightarrow [[q]]s',h').$$

Now, we can use Hoare triples to give rules of reasoning for every individual command of the language $\mathcal{L}$. It is sometimes also called the *axiomatic semantics* of the language. These rules in a way give an alternative semantics to the commands.

Axiomatic Semantics: Consider the following set of axioms (AXIOMS-I) and structural rules for reasoning with commands, that does not use pointers. Here,

$Q[e/x]$ represents the proposition Q with every free occurrence of x replaced by the expression $e, Mod(c)$ represents the set of variables modified by c, and $Free(R)$ represents the set of free variables in R.

AXIOMS-I

$$\text{skip}\ \forall P : \texttt{Assert}, \{P\}\texttt{skip}\{P\}$$

$$\text{assign}\ \frac{}{\{Q[e/x]\}\texttt{x} := \texttt{e}\{Q\}}$$

$$\text{seq}\ \frac{\{P\}c_1\{Q\} \quad \{Q\}c_2\{R\}}{\{P\}c_1; c_2\{R\}}$$

$$\text{if}\ \frac{\{P \wedge e\}c_1\{Q\} \quad \{P \wedge \neg e\}c_2\{Q\}}{\{P\}\texttt{if}\ e\ \texttt{then}\ c_1\ \texttt{else}\ c_2\{Q\}}$$

$$\text{while}\ \frac{\{I \wedge e\}c\{I\}}{\{I\}\texttt{while}\ e\ \texttt{do}\ c\{I \wedge \neg e\}}$$

STRUCTURAL RULES

$$\text{conseq}\ \frac{P \Rightarrow P'\{P'\}c\{Q'\}Q' \Rightarrow Q}{\{P\}c\{Q\}}$$

$$\text{extractE}\ \frac{\{P\}c\{Q\}}{\{\exists x.P\}c\{\exists x.Q\}}\ x \notin Free(c)$$

$$\text{var-sub}\ \frac{\{P\}c\{Q\}}{(\{P\}c\{Q\})[E_1/x_1, \ldots, E_k/x_k]} \quad \begin{array}{l} x_i \in Mod(c)\ \text{implies} \\ \forall\ j \neq i,\ E_i \notin Free(E_j) \end{array}$$

$$\text{constancy}\ \frac{\{P\}c\{Q\}}{\{P \wedge R\}c\{Q \wedge R\}}\ Mod(c) \cap Free(R) = \phi$$

- For the subset of language $\mathcal{L}$, that does not use pointers, these axioms and structural rules are known to be sound as well as complete [3] with respect to the operational semantics.
- The benefit of using these axioms is that we can work on a more abstract level specifying and proving program correctness in an axiomatic way without bothering about low-level details of states.
- Note that the last four commands of the language $\mathcal{L}$, which manipulate pointers, are different from the normal variable assignment command. Their right hand side is not an expression. Therefore, the assign rule of Hoare logic is not applicable to them.

Array Revisited: Now, let us go back to the same array assignment problem which we discussed in the introduction. Let Q be the postcondition for the command `a[i]:= 8`, where `a[i]` refers to the ith element of array.

- In our language $\mathcal{L}$, this is the same as the command `[a+i]:= 8`; however, for convenience, we will continue with the usual notation of arrays.

The command `a[i]:= 8` looks similar to the variable assignment command. But, we cannot apply Hoare assign rule to get Q[8/a`a[i]`] as weakest precondition. One should not treat `a[i]` as a local variable, because the assertion Q may contain references such as `a[j]` that may or may not refer to `a[i]`. Instead, we can model the above command as `a := update(a, i, 8)`, where `update(a,i,8)` $[i] =$ 8 and `update(a,i,8)` $[j] = \mathrm{a}\,[j]$ for $j \neq i$.

- The effect of executing `a[i]:= v` is same as assigning variable a an altogether new array value "`update (a,i,v)`".
- In this way, a is acting like a normal variable; hence, we have the following rule for array assignment,

$$\text{array} - \text{assign}\,\frac{}{\{Q[\mathtt{update}(\mathtt{a},\mathtt{i},\mathtt{v})/\mathtt{a}]\}\mathtt{a}[\mathtt{i}] := \mathtt{v}\{Q\}}$$

Let us try to prove the following specification using the above rule. $\{i \neq j \wedge a[i] = 4 \wedge a[j] = 4\}\ \mathtt{a}[\mathtt{i}] := 8\ \{a[i] = 8 \wedge a[j] = 4 \wedge i \neq j\}$..

Let $P = \{i \neq j \wedge a[i] = 4 \wedge a[j] = 4\}$ and $Q = \{a[i] = 8 \wedge a[j] = 4 \wedge i \neq j\}$.

Then we have, Q [update(a,i,8)/a]

$$\begin{aligned}
&=\{\mathtt{update}(\mathtt{a},\mathtt{i},8)[i] = 8 \wedge \mathtt{update}(\mathtt{a},\mathtt{i},8)[j] = 4 \wedge i \neq j\}\\
&=\{8 = 8 \wedge a[j] = 4 \wedge i \neq j\}\\
&=\{a[j] = 4 \wedge i \neq j\}
\end{aligned}$$

Thus,

$$\frac{P \Rightarrow Q[\mathtt{update}(\mathtt{a},\mathtt{i},8)/\mathtt{a}] \quad \{Q[\mathtt{update}(\mathtt{a},\mathtt{i},8)/\mathtt{a}]\}\mathtt{a}[\mathtt{i}] := 8\ \{Q\}}{\{P\}\mathtt{a}[\mathtt{i}] := 8\{Q\}}$$

Hence, we have a correct rule for deducing valid specifications about array assignments. However, the approach looks very clumsy.

- We still need to fill in all minute details of index disjointness in the specification.
- Moreover, it seems very artificial to interpret a local update to an array cell as a global update to the whole array. At least, it is not the programmer's way of understanding an array element update.

- The idea of separation logic is to embed the principle of such local actions in the separating conjunction. It helps in keeping the specifications succinct by avoiding the explicit mention of memory disjointness.

3 New Assertions and Inference Rules

In this section, we present the axioms corresponding to the pointer-manipulating commands. The set of assertions, which we use for this purpose, goes beyond the predicates used in Hoare logic. Following is the syntax of the new assertions:

$$
\begin{aligned}
\langle assert \rangle \ ::= \ & \ldots \\
\mid \ & \mathbf{emp} \\
\mid \ & \langle aexp \rangle \mapsto \langle aexp \rangle \\
\mid \ & \langle assert \rangle * \langle assert \rangle \\
\mid \ & \langle assert \rangle -\!\!* \langle assert \rangle
\end{aligned}
$$

It is important to note that the meaning of these new assertions depend on both the store and the heap.

- **emp**
  The heap is empty.
- $e_1 \mapsto e_2$
  The heap contains a single cell, at address e_1 with contents e_2.
- $p_1 * p_2$
  The heap can be split into two disjoint parts in which p_1 and p_2 hold, respectively.
- $p_1 - *p_2$
  If the current heap is extended with a disjoint part in which p_1 holds, then p_2 holds for the extended heap.

For convenience, we introduce some more notations for the following assertions:

$$e \mapsto - \triangleq \exists x. e \mapsto x \text{ where } x \text{ is not free in } e$$

$$e \hookrightarrow e' \triangleq e \mapsto e' * \mathbf{true}$$

$$e \mapsto e_1, \ldots, e_n \triangleq e \mapsto e_1 * \ldots * e + n - 1 \mapsto e_n$$

$$e \hookrightarrow e_1, \ldots, e_n \triangleq e \hookrightarrow e_1 * \ldots * e + n - 1 \hookrightarrow e_n$$

$$\text{iff } e \mapsto e_1, \ldots, e_n * \mathbf{true}$$

We now consider a simple example to explore some of the interesting features of separating conjunction. Let $h_1 = \{(sx, 1)\}$ and $h_2 = \{(sy, 2)\}$ be heaps where s is a store such that $sx \neq sy$. Then, one can verify the following

1.	$[\![x \mapsto 1 * y \mapsto 2]\!] s\, h$	iff	$h = h_1.h_2$
2.	$[\![x \mapsto 1 \wedge x \mapsto 1]\!] s\, h$	iff	$h = h_1$
3.	$[\![x \mapsto 1 * x \mapsto 1]\!] s\, h$	iff	**false**
4.	$[\![x \mapsto 1 \vee y \mapsto 2]\!] s\, h$	iff	$h = h_1$ or $h = h_2$
5.	$[\![x \mapsto 1 * (x \mapsto 1 \vee y \mapsto 2)]\!] s\, h$	iff	$h = h_1.h_2$
6.	$[\![(x \mapsto 1 \vee y \mapsto 2) * (x \mapsto 1 \vee y \mapsto 2)]\!] s\, h$	iff	$h = h_1.h_2$
7.	$[\![(x \mapsto 1 * y \mapsto 2 * (x \mapsto 1 \vee y \mapsto 2)]\!] s\, h$	iff	**false**

Assertions 2 and 3 illustrate the difference between the behavior of classical conjunction and separating conjunction. Both the occurrences of $x \mapsto 1$ in the assertion $[\![x \mapsto 1 * x \mapsto 1]\!] s\, h$ are true for the same singleton heap h_1. Hence, any heap h can never be split into two disjoint parts that satisfies $x \mapsto 1$. One can also compare assertions 6 and 7, which look similar in structure, but have different behaviors.

The separating conjunction obeys commutative, associative and some distributive as well as semi-distributive laws. The assertion **emp** behaves like a neutral element. Most of these properties are contained in the following axiom schema. Note the use of uni-directional implications in $(p_1 \wedge p_2) * q \Rightarrow (p_1 * q) \wedge (p_2 * q)$ and $(\forall x \cdot p) * q \Rightarrow \forall x \cdot (p * q)$.

$$
\begin{aligned}
p * \mathbf{emp} &\Leftrightarrow p \\
p_1 * p_2 &\Leftrightarrow p_2 * p_1 \\
(p_1 * p_2) * p_3 &\Leftrightarrow p_1 * (p_2 * p_3) \\
(p_1 \vee p_2) * q &\Leftrightarrow (p_1 * q) \vee (p_2 * q) \\
(p_1 \wedge p_2) * q &\Rightarrow (p_1 * q) \wedge (p_2 * q) \\
(\exists x \cdot p) * q &\Leftrightarrow \exists x \cdot (p * q) \text{ where } x \text{ is not free in } q \\
(\forall x \cdot p) * q &\Rightarrow \forall x \cdot (p * q) \text{ where } x \text{ is not free in q}
\end{aligned}
$$

New Axioms for pointers: The axioms needed to reason about pointers are given below. There is one axiom for every individual command.

AXIOMS-II

$$\text{alloc}\ \frac{}{\{x = X \wedge \mathbf{emp}\}\texttt{x} := \texttt{cons}(e_1, \ldots, e_k)\{x \mapsto e_1[X/x], \ldots, e_k[X/x]\}}$$

$$\text{lookup}\ \frac{}{\{e \mapsto v \wedge x = X\}\texttt{x} := [\texttt{e}]\{x = v \wedge e[X/x] \mapsto v\}}$$

$$\text{mut}\ \frac{}{\{e \mapsto -\}[\texttt{e}] := \text{e}'\{e \mapsto e'\}}$$

$$\text{free}\ \frac{}{\{e \mapsto -\}\texttt{free}\,(e)\{\mathbf{emp}\}}$$

Allocate	The first axiom, called *alloc*, uses variable X in its precondition to record the value of x before the command is executed. It says that if execution begins with empty heap and a store with $x = X$, then it ends with k contiguous heap cells having appropriate values
Lookup	The second axiom, called *lookup*, again uses X to refer to the value of x before execution. It asserts that the content of heap is unaltered. The only change is in the store where the new value of x is modified to the value at old location e
Mutate	The third axiom, called *mut*, says that if e points to something beforehand, then it points to e' afterward. This resembles the natural semantics of Mutation
Free	The last axiom, called *free*, says that if e is the only allocated memory cell before execution of the command, then in the resulting state there will be no active cell. Note that, the singleton heap assertion is necessary in the precondition to assure **emp** in the postcondition
Frame	The last rule among the structural rules, called *frame*, says that one can extend local specifications to include any arbitrary claims about variables and heap segments which are not modified or mutated by c. The frame rule can be thought as a replacement to the rule of constancy when pointers are involved

STRUCTURAL RULES-II

$$\text{conseq}\frac{P \Rightarrow P' \quad \{P'\}c\{Q'\} \quad Q' \Rightarrow Q}{\{P\}c\{Q\}}$$

$$\text{extractE}\frac{\{P\}c\{Q\}}{\{\exists x.P\}c\{\exists x.Q\}} x \notin Free(c)$$

$$\text{var-sub}\frac{\{P\}c\{Q\}}{(\{P\}c\{Q\})[E_1/x_1, \ldots, E_k/x_k]} \quad \begin{array}{l} x_i \in Mod(c) \text{ implies} \\ \forall j \neq i,\ E_i \notin Free(E_j) \end{array}$$

$$\text{frame}\frac{\{p\}c\{q\}}{\{p * r\}c\{q * r\}} Mod(c) \cap Free(r) = \phi$$

- Note that the expressions are intentionally kept free from the `cons` and [−] operators. The reason for this restriction is that the power of the above proof system strongly depends upon the ability to use expressions in place of variables in an assertion.
- In particular, a tautology should remain a valid assertion on replacing variables with expressions.

- If we could substitute $\mathtt{cons}\ (e_1, e_2)$ for x in the tautology $x = x$, we obtain $\mathtt{cons}\ (e_1, e_2) = \mathtt{cons}\ (e_1, e_2)$, which may not be a valid assertion if we wish to distinguish between different addresses having the same content.
- Similarly $[-]$ cannot be used in expressions because of the way it interacts with separating conjunction. For example, consider substituting $[e]$ for x and y in the tautology $x = x * y = y$. Clearly, $[e] = [e] * [e] = [e]$ is not a valid assertion.
- Note that each axiom mentions only the portion of heap accessed by the corresponding command. In this sense, the axioms are local. Hence, a separate rule, called frame rule, is needed to extend this local reasoning to a global context.
- These axioms can easily be proved to be sound with respect to the operational semantics of the language $\mathcal{L}$.
- Moreover, Yang in his thesis [4] has shown that all valid Hoare triples can be derived using the above collection of axioms and the structural rules. In this sense, these set of axioms and structural rules are also complete.

Derived Rules: Although the small set of rules discussed so far is complete, it is not practical. Proving a specification using this small set of axioms often requires extensive invocations of the structural rules. Therefore, it is good to have some derived rules that can be applied at once in common situations. We now list some other useful rules that can be derived from the natural semantics of the language $\mathcal{L}$. A more detailed discussion about these rules can be found in [5]. Note that x, x' and X are all distinct variables.

- Assignment
  - Forward reasoning

$$\frac{}{\{x = X\}\mathsf{x} := \mathsf{e}\{x = e[X/x]\}}$$

  - Floyd's forward running axiom

$$\frac{}{\{P\}\mathsf{x} := \mathsf{e}\{\exists x'.x = e[x'/x] \wedge P[x'/x]\}}$$

- Mutation
  - Global reasoning

$$\frac{}{\{(e \mapsto -) * r\}[\mathsf{e}] := \mathsf{e'}\{(e \mapsto e') * r\}}$$

  - Backward reasoning

$$\frac{}{\{(e \mapsto -) * ((e \mapsto e') -\!* p)\}[\mathsf{e}] := \mathsf{e'}\{p\}}$$

- Free
  - Global (backward) reasoning

$$\overline{\{(e \mapsto -) * r\}\texttt{free (e)}\{r\}}$$

- Allocation
  - Global reasoning (forward)

$$\overline{\{r\}\texttt{x := cons}(e_1,\ldots,e_k)\{\exists x'.(x \mapsto e_1[x'/x],\ldots,e_k[x'/x]) * r[x'/x]\}}$$

  - Backward reasoning

$$\overline{\{\forall x'.(x' \mapsto e_1,\ldots,e_k) - *p[x'/x]\}\texttt{x := cons}(e_1,\ldots,e_k)\{p\}}$$

- Lookup
  - Global reasoning

$$\overline{\{\exists x''.(e \mapsto x'') * r[x/x']\}\texttt{x := [e]}\{\exists x'.(e[x'/x] \mapsto x) * r[x/x'']\}}$$

    Here, x, x' and x'' are distinct, x' and x'' do not occur free in e, and x is not free in r.
  - Backward reasoning

$$\overline{\{\exists x'.(e \mapsto x') * ((e \mapsto x') - *p[x'/x])\}\texttt{x := [e]}\{p\}}$$

4 Annotated Proofs

Proof outlines: We have already used assertions in Hoare triples to state what is true before and after the execution of an instruction. In a similar way, an assertion can also be inserted between any two commands of a program to state what must be true at that point of execution. Placing assertions in this way is also called *annotating* the program.

For example, consider the following annotated program that swaps the value of variable x and y using a third variable z. Note the use of X and Y to represent the initial values of variable x and y, respectively.

$$\{x = X \wedge y = Y\}$$
$$z := x;$$
$$\{z = X \wedge x = X \wedge y = Y\}$$
$$x := y;$$
$$\{z = X \wedge x = Y \wedge y = Y\}$$
$$y := z;$$
$$\{x = Y \wedge y = X\}$$

Validity of each Hoare triple in the above program can easily be checked using axioms for assignment. Hence, one can conclude that the program satisfies its specification.

- A program together with an assertion between each pair of statements is called a *fully annotated* program.
- One can prove that a program satisfies its specification by proving the validity of every consecutive Hoare triple which is present in its annotated version. Hence, a fully annotated program provides a complete *proof outline* for the program.

Now, we consider another annotated program that involves assertions from the separation logic. Note that the assertion $(x \mapsto a, o) * (x + o \mapsto b, -o)$ can be used to describe a circular offset list. Here is a sequence of commands that creates such a cyclic structure:

$$1\ \{\mathbf{emp}\}$$
$$x := \texttt{cons}\ (a, a);$$
$$2\ \{x \mapsto a, a\}$$
$$t := \texttt{cons}\ (b, b);$$
$$3\ \{(x \mapsto a, a)\ *\ (t \mapsto b, b)\}$$
$$[x + 1] := t - x;$$
$$4\ \{(x \mapsto a, t - x)\ *\ (t \mapsto b, b)\}$$
$$[t + 1] := x - t;$$
$$5\ \{(x \mapsto a, t - x)\ *\ (t \mapsto b, x - t)\}$$
$$6\ \{\exists o.(x \mapsto a, o) * (x + o \mapsto b, -o)\}$$

The above proof outline illustrates two important points.

- First, a label is used against each assertion so that referring becomes easy in the future discussions.
- Secondly, the adjacent assertions—e.g., here the assertions 5 and 6—mean that the first implies the second.

Also, note the use of * in assertion 3. It ensures that $x+1$ is different from t and $t+1$, and hence, the assignment $[x+1] := t - x$ cannot affect the $t \mapsto b, b$ clause. A similar reasoning applies to the last command as well.

Inductive definitions: When reasoning about programs which manipulate data structures, we often need to use inductively defined predicates describing such structures. For example, in any formal setting, if we wish to reason about the contents of a linked list, we would like to relate it to the abstract mathematical notion of sequences.

- Consider the following inductive definition of a predicate that describes the content of a linked list

$$\texttt{listrep}\,\epsilon(i,j) \triangleq i = j \wedge \mathbf{emp}$$
$$\texttt{listrep}\,\mathrm{a}.\alpha(i,k) \triangleq i \neq j \wedge \exists j.i \mapsto \mathrm{a}, j * \texttt{listrep}\,\alpha(j,k)$$

Here, α denotes a mathematical sequence. Informally, the predicate $\texttt{listrep}\,\alpha(x,y)$ claims that x points to a linked list segment ending at y and the contents (head elements) of that segment are the sequence α.

- While proving programs in this section, we use $x \overset{\alpha}{\rightsquigarrow} y$ as an abbreviation for $\texttt{listrep}\,\alpha(x,y)$ and $\alpha^{\dagger}$ to represent the reverse of the sequence α.

Proof of In-place list reversal: Consider the following piece of code that performs an in-place reversal of a linked list:

$\{i \overset{\alpha_0}{\rightsquigarrow} \boxtimes\}$

$/ *$ i points to the initial linked list $* /$

$\mathtt{j} := \boxtimes;$

$\mathtt{while}\,\mathrm{i} \neq \boxtimes\,\mathtt{do}$

$(\mathtt{k} := [\mathtt{i}+1];\ [\mathtt{i}+1] := \mathtt{j};\ \mathtt{j} := \mathtt{i};\ \mathtt{i} := \mathtt{k};)$

$/ *$ j points to the in place reversal of the initial list pointed by i $* /$

$\{j \overset{\alpha_0^{\dagger}}{\rightsquigarrow})\boxtimes\}$

Here, ⊠ represents the null pointer. On a careful analysis of the code, it is easy to see that:

- At any iteration of the while loop, variables i and j point to two different list segments having the contents α and β such that concatenating β at the end of $\alpha^{\dagger}$ will always result in α_0.
- Thus, we have the following loop invariant

$$\exists\alpha,\beta\cdot(\mathrm{i}\overset{\alpha}{\rightsquigarrow}\boxtimes*\mathrm{j}\overset{\beta}{\rightsquigarrow}\boxtimes\wedge\alpha_0^{\dagger}=\alpha^{\dagger}\cdot\beta$$

where α_0 represents the initial content of linked list pointed by variable i.

Also, note the use of separating conjunction in the loop invariant instead of the usual classical conjunction. If there is any sharing between the lists i and j, then the program may malfunction. The use of a classical conjunction here cannot guarantee such non-sharing.

It is easy to see how the postcondition of the list reversal program follows from the above loop invariant. The following sequence of specifications gives a derivation of the postcondition assuming the loop invariant and the termination condition $i = \boxtimes$,

$$8\,\{\exists\alpha,\beta.(\mathrm{i}\overset{\alpha}{\rightsquigarrow}\boxtimes*\mathrm{j}\overset{\beta}{\rightsquigarrow}\boxtimes)\wedge\alpha_0^{\dagger}=\alpha^{\dagger}.\beta\wedge i=\boxtimes\}$$

$$8\mathrm{a}\,\{\exists\beta.(\mathrm{i}\overset{\epsilon}{\rightsquigarrow}\boxtimes*\mathrm{j}\overset{\beta}{\rightsquigarrow}\boxtimes)\wedge\alpha_0^{\dagger}=\epsilon^{\dagger}.\beta\}$$

$$8\mathrm{b}\,\{\exists\beta.(\mathrm{i}\overset{\epsilon}{\rightsquigarrow}\boxtimes*\mathrm{j}\overset{\beta}{\rightsquigarrow}\boxtimes)\wedge\alpha_0^{\dagger}=\beta\}$$

$$8\mathrm{c}\,\{(\mathrm{i}\overset{\epsilon}{\rightsquigarrow}\boxtimes*\mathrm{j}\overset{\alpha_0^{\dagger}}{\rightsquigarrow}\boxtimes)\}$$

$$8\mathrm{d}\,\{(\mathrm{j}\overset{\alpha_0^{\dagger}}{\rightsquigarrow}\boxtimes)\}$$

where the loop invariant can be verified using the following proof outline:

$$1\,\{\exists\alpha,\beta.(i\overset{\alpha}{\rightsquigarrow}\boxtimes*j\overset{\beta}{\rightsquigarrow}\boxtimes)\wedge\alpha_0^{\dagger}=\alpha^{\dagger}.\beta\wedge i\neq\boxtimes\}$$

$$2\{\exists\alpha'.(i\mapsto a,p*p\overset{\alpha'}{\rightsquigarrow}\boxtimes*j\overset{\beta}{\rightsquigarrow}\boxtimes)\wedge\alpha_0^{\dagger}=(a.\alpha')^{\dagger}.\beta\}$$

$\mathtt{k} := [\mathtt{i}+1];$

$$3\,\{\exists\alpha'.(i\mapsto a,k*k\overset{\alpha'}{\rightsquigarrow}\boxtimes*j\overset{\beta}{\rightsquigarrow}\boxtimes)\wedge\alpha_0^{\dagger}=(a.\alpha')^{\dagger}.\beta\}$$

$[\mathtt{i}+1] := \mathtt{j};$

$$4\,\{\exists\alpha'.(i\mapsto a,j*k\overset{\alpha'}{\rightsquigarrow}\boxtimes*j\overset{\beta}{\rightsquigarrow}\boxtimes)\wedge\alpha_0^{\dagger}=(a.\alpha')^{\dagger}.\beta\}$$

$$5\,\{\exists\alpha',\beta'.(k\overset{\alpha'}{\rightsquigarrow}\boxtimes*i\overset{\beta'}{\rightsquigarrow}\boxtimes)\wedge\alpha_0^{\dagger}=(\alpha')^{\dagger}.\beta'\}$$

$\mathtt{j} := \mathtt{i};$

$$6\,\{\exists\alpha',\beta'.(k\overset{\alpha'}{\rightsquigarrow}\boxtimes*j\overset{\beta'}{\rightsquigarrow}\boxtimes)\wedge\alpha_0^{\dagger}=(\alpha')^{\dagger}.\beta'\}$$

$\mathtt{i} := \mathtt{k};$

$$7\,\{\exists\alpha',\beta'.(i\overset{\alpha'}{\rightsquigarrow}\boxtimes*j\overset{\beta'}{\rightsquigarrow}\boxtimes)\wedge\alpha_0^{\dagger}=(\alpha')^{\dagger}.\beta'\}$$

$$7\mathrm{a}\,\{\exists\alpha,\beta.(i\overset{\alpha}{\rightsquigarrow}\boxtimes*j\overset{\beta}{\rightsquigarrow}\boxtimes)\wedge\alpha_0^{\dagger}=(\alpha)^{\dagger}.\beta\}$$

Moreover, the following sequence of assertions gives a detailed proof of the implications 1 $\Longrightarrow$ 2 and 4 $\Longrightarrow$ 5:

$$1\ \{\exists\alpha,\beta.(i \overset{\alpha}{\rightsquigarrow} \boxtimes * j \overset{\beta}{\rightsquigarrow} \boxtimes) \wedge \alpha_0^{\dagger} = \alpha^{\dagger}.\beta \wedge i \neq \boxtimes\}$$
$$1\text{a}\ \{\exists a,\alpha',\beta.(i \overset{a.\alpha'}{\rightsquigarrow} \boxtimes * j \overset{\beta}{\rightsquigarrow} \boxtimes) \wedge \alpha_0^{\dagger} = (a.\alpha')^{\dagger}.\beta\}$$
$$1\text{b}\ \{\exists a,\alpha',\beta,p.(i \mapsto a,p * p \overset{\alpha'}{\rightsquigarrow} \boxtimes * j \overset{\beta}{\rightsquigarrow} \boxtimes) \wedge \alpha_0^{\dagger} = (a.\alpha')^{\dagger}.\beta\}$$
$$2\ \{\exists\alpha'.(i \mapsto a,p * p \overset{\alpha'}{\rightsquigarrow} \boxtimes * j \overset{\beta}{\rightsquigarrow} \boxtimes) \wedge \alpha_0^{\dagger} = (a.\alpha')^{\dagger}.\beta\}$$
$$4\ \{\exists\alpha'.(i \mapsto a,j * k \overset{\alpha'}{\rightsquigarrow} \boxtimes * j \overset{\beta}{\rightsquigarrow} \boxtimes) \wedge \alpha_0^{\dagger} = (a.\alpha')^{\dagger}.\beta\}$$
$$4\text{a}\{\exists\alpha'.(k \overset{\alpha'}{\rightsquigarrow} \boxtimes * i \mapsto a,j * j \overset{\beta}{\rightsquigarrow} \boxtimes) \wedge \alpha_0^{\dagger} = (a.\alpha')^{\dagger}.\beta\}$$
$$4\text{b}\ \{\exists\alpha'.(k \overset{\alpha'}{\rightsquigarrow} \boxtimes * i \overset{\alpha.\beta}{\rightsquigarrow} \boxtimes) \wedge \alpha_0^{\dagger} = (\alpha')^{\dagger}.a.\beta\}$$
$$5\ \{\exists\alpha',\beta'.(k \overset{\alpha'}{\rightsquigarrow} \boxtimes * i \overset{\beta'}{\rightsquigarrow} \boxtimes) \wedge \alpha_0^{\dagger} = (\alpha')^{\dagger}.\beta'\}$$

– Explanations:
  - Most of the proof steps, especially those around a command, comprise Hoare triples, which can easily be verified using the axioms for the corresponding commands.
  - Note the use of * instead of $\wedge$ in assertion 3. It ensures that $i+1$ is different from k and j. Hence, an attempt to mutate the location $i+1$ does not affect the remaining two clauses $k \overset{\alpha'}{\rightsquigarrow} \boxtimes$ and $j \overset{\beta}{\rightsquigarrow} \boxtimes$.
  - We can obtain 1a from 1 by using definition of $i \overset{\alpha}{\rightsquigarrow} \boxtimes$ with the fact that $i \neq \boxtimes$. Then, we unfold the definition of $i \overset{\alpha.\alpha'}{\rightsquigarrow} \boxtimes$ to obtain 1b from 1a. Finally, instantiating $\exists a$ takes us to 2.
  - 4a is a simple rearrangement of 4. Since $i \overset{\alpha.\beta}{\rightsquigarrow} \boxtimes$ is a shorthand for $i \mapsto a,j * j \overset{\beta}{\rightsquigarrow} \boxtimes$, we can obtain 4b from 4a. Finally, we obtain 5 by generalizing $a.\beta$ in 4b as β' using the existential quantifier.

5 Conclusion

In this article, we reviewed some of the important features of separation logic, that first appeared in [1, 6, 7]. We illustrated the difficulties that arise in the reasoning and development of modular programs due to unwanted aliasing arising due to pointer data structures. We introduced the notion of separating conjunction as a tool to deal with it and presented separation logic as an extension to Hoare Logic using a programming language with four essential pointer-manipulating commands. These commands performed heap operations such as lookup, update, allocation, and de-allocation. The new set of assertions and axioms of separation logic were presented, and examples were given to illustrate the unique features of these new assertions and axioms. Using the axioms of separation logic, we also saw illustrations of proofs of some real and non-trivial programs such as in-place list reversal programs.

The key idea of separating conjunction was inspired by Burstall's [8] "distinct non-repeating tree systems". It is based on the idea of organizing assertions to localize the effect of a mutation. The separating conjunction gives us a succinct and more intuitive way to describe memory disjointness when pointers are involved. However, it is not the only possible way. One can see [9] for references and other papers on proving pointer programs using standard Hoare Logic.

In this paper, we considered simple data structures to illustrate the power of separation logic. A more elaborate discussion with a variety of data structures can be found in [1, 5]. Reasoning becomes difficult when data structures use more sharing. In this direction, one can refer Yang's proof [10] of the Schorr-Waite graph marking algorithm.

We did not talk much about the proof theory behind separation logic. For a detailed discussion on the soundness and completeness results, one can refer [4, 11]. The soundness results for most of the derived rules, presented in this paper, can also be found in [5].

Finally, it should be noted that the challenge here was to present a formalism that captured certain key intuitions embodied in the informal local reasoning employed by astute programmers. Programmers often assume non-sharing between data structures, which needs explicit mention when using standard techniques such as Hoare Logic. On the other hand, memory disjointness is default in the separating conjunction. Hence, separation logic gives us a more natural and concise way to model a programmer's reasoning.

References

1. J.C. Reynolds, Separation logic: a logic for shared mutable data structures, in *Proceedings Seventeenth Annual IEEE Symposium on Logic in Computer Science* (Los Alamitos, California, IEEE Computer Society, 2002) pp. 55–74
2. C.A.R. Hoare, An axiomatic basis for computer programming. Commun. ACM. **12**(10):576–580, 583 (1969)
3. G. Winskel, *The Formal Semantics of Programming Languages* (MIT Press, USA, 1993)
4. H. Yang, *Local Reasoning for Stateful Programs*. Ph. D. dissertation (University of Illinois, Urbana-Champaign, Illinois, 2001)
5. J.C. Reynolds, *An Introduction to Separation Logic (Preliminary Draft)*. http://www.cs.cmu.edu/jcr/copenhagen08.pdf. 23 Oct 2008
6. J.C. Reynolds, Intuitionistic reasoning about shared mutable data structures, in *Millennial Perspectives in Computer Science*, ed. by J. Davies, B. Roscoe, J. Woodcock (Houndsmill, Hampshire, Palgrave, 2000) pp. 303–321

7. P.W. O'Hearn, J.C. Reynolds, H. Yang, Local reasoning about programs that alter data structures, in *Proceedings of 15th Annual Conference of the European Association for Computer Science Logic: CSL 2001,* Lecture Notes in Computer Science (Springer, Berlin, 2001)
8. R.M. Burstall, Some techniques for proving correctness of programs which alter data structures, in *Machine Intelligence*, ed. by B. Meltzer, D. Michie, vol 7 (Edinburgh University Press, Edinburgh, Scotland, 1972), pp 23–50
9. R. Bornat, Proving pointer programs in Hoare logic. *Math. Program Constr.* (2000)
10. H. Yang, An example of local reasoning in BI pointer logic: the Schorr-Waite graph marking algorithm, in *SPACE 2001: Informal Proceedings of Workshop on Semantics, Program Analysis and Computing Environments for Memory Management*, ed by F. Henglein, J. Hughes, H. Makholm, H. Niss (IT University of Copenhagen, Denmark, 2001), pp. 41–68
11. Peter W. O'Hearn, David J. Pym, The logic of bunched implications. Bull. Symbol. Logic **5** (2), 215–244 (1999)

mDSM: A Transformative Approach to Enterprise Software Systems Evolution

David Threm, Liguo Yu, S.D. Sudarsan and Srini Ramaswamy

Abstract The engineering of enterprise software systems suffers from an inherent lack of creativity and innovation and is often left to user-centric incremental changes that are not often disruptive enough for business needs. A design-driven approach to systems creates opportunities for transformative evolution of such systems that are both immediate and futuristic in their impact. Software systems stability can be maintained and monitored during evolution utilizing architectural-level, program-level, and information-level stability metrics. Despite increasing complexities involved in the design, development, and testing of such large-scale software systems, they are often predicated by simple techniques for decomposition, generalization, and specification. However, as always they are much more difficult to merge back together in order to rationalize the entire architecture for the levels of confidence necessary during testing, deployment, and commissioning of these systems. mDSM, an extension to Design Structure Matrix

Note: The earlier results of this study (Sects. 2.4, 3.1, and 4.1) were published in Proceedings of the Twenty-sixth IEEE International Symposium on Software Reliability Engineering, Gaithersburg, MD, November 2–5, 2015.

D. Threm
University of Arkansas—Little Rock, Little Rock, AR, USA
e-mail: dsthrem@ualr.edu

L. Yu
Computer Science and Informatics, Indiana University South Bend, South Bend, IN, USA
e-mail: ligyu@iusb.edu

S.D. Sudarsan
Industrial Software Systems, ABB Corporate Research Center, Bangalore, India
e-mail: sdsudarsan@gmail.com

S. Ramaswamy (✉)
ABB Inc., Cleveland, OH, USA
e-mail: srini@ieee.org

H. Mohanty et al. (eds.), *Trends in Software Testing*,
DOI 10.1007/978-981-10-1415-4_6

(DSM) approach to software systems design and testing, is a methodology developed by the authors to address design-driven rationalization of such complex software system architectures.

Keywords Software testing · Software design · Stability · Architecture · Evolution · Rationalization

1 Introduction

> From a very early age, we are taught to break apart problems, to fragment the world. This apparently makes complex tasks and subjects more manageable, but we pay a hidden, enormous price… after a while, we give up trying to see the whole altogether - Senge

Design-driven innovation is the radical innovation of meaning [1]. Radical innovation of meaning is how quickly the technology is capable of change. User-driven, incremental change is simply not disruptive enough for many organizations. The Software Design and Test communities need tools that allow them to be disruptive, while not losing sight of the bigger picture; we have to be able to see the forest through the trees. As the quote above by Senge implies [2], we are very good at the decomposition, or breaking apart, of systems, but we tend to lose our vision of how design and testing of systems impacts the big picture. If software systems are to be truly disruptive, innovative, transformational, and market-place differentiators, we have to be able to quickly rationalize architectural-, program- and information-level artifacts of systems in the design and test phases of software development life cycle without losing the larger picture.

Current approaches to measuring software design and testing quality in large-scale software systems heavily rely on metrics. Established metrics calculate burn down rates based on bug types, classes, design issues, severity, and similar classifications. However, these classifications do not focus on whether the software system performs with improved stability over time or how systems or subsystems changes impact design and testing rationalization.

The mDSM, a metrics-based extension to the DSM, allows us to retain the larger picture. By utilizing the mDSM with evolutionary stability metrics, we can retain our view of a software system in its entirety without losing sight of how the modules, units, subsystems, or components interact. We can use evolutionary stability metrics to evaluate a software system's evolution at the lowest level, and the software system can still be rationalized in totality.

While the importance of systems rationalization is tantamount, we cannot underestimate the importance of scalability and agility to overall design and testing. From a scalability perspective, we have to be able to create or evolve systems that are capable of acquiring or divesting other systems. Scalable systems are capable of scaling up and down, whatever the organization or the customers require. Agility plays a role in just being "good enough" and flexible to organizational needs. The software system needs to adapt, but only at the appropriate amount for timeliness and completeness of solution.

This chapter is arranged as follows: Sect. 2—Background concepts on software design, testing, interaction of design and testing, software stability, and the design structure matrix (DSM); Sect. 3—mDSM and Evolutionary stability; Sect. 4—Case Studies; Sect. 5—Advantages of mDSM and Design and Testing Rationalization; and Sect. 6—Summary and Conclusions.

2 Background Concepts

2.1 Software Design

Software design is the process that transforms a specification into a structure suitable for implementation [3]. The two major stages of software design are architectural design and detailed design [4]. In the architectural design stage, the system is decomposed into a unit-, module- or component-based systems model. The detailed design phase utilizes the artifacts created by the architectural design phase to create a procedural representation of the system.

Architectural design techniques are further broken down to process-oriented design and data-oriented design techniques [4]. Process-oriented techniques focus on functional decomposition and structural software architecture. Data-oriented techniques focus on data structure and flow.

Detailed design techniques include graphical representation techniques and language representation techniques [4]. Graphical representation techniques include graphs and flowcharts. Language representation techniques include program description languages, PDLs.

Unified Modeling Language (UML), a model-driven architecture (MDA), is a software design methodology that can be either proprietary or open source [5]. UML models can be static or structural view as well as dynamic view of the systems behavior. The UML methodology tends to blur the lines between architectural and detail design stages. Traditional development processes are also impacted, since many models can also generate the source code base of a software system. While UML may be viewed as the ideal, its use in software development practice has been questioned. UML in practical use is simply used for modeling purposes [6].

As Yu, Threm, and Ramaswamy noted in [7], well-designed software systems exhibit high cohesion or interaction within a module and low coupling between modules. Figure 1 accurately depicts a software system with high cohesion within components and low coupling between larger component systems. A software system is made up of components: C1, C2, and C3 and exhibits low coupling and contain subcomponents C11–C32, which exhibit high cohesion within their respective components.

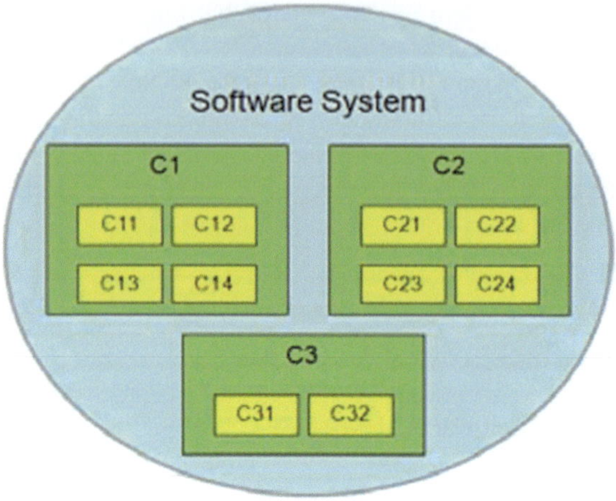

Fig. 1 High cohesion and low coupling

2.2 *Software Testing*

Software testing is a verification and validation technique that ensures software is developed to meet both its specification and its users' needs [3]. Simply stated, software testing is used to ensure software quality. The four generally accepted testing methods during the software development life cycle are summarized in Table 1. Defects and errors are expected to be detected during unit, integration, and system testing. Acceptance testing, often referred to as user acceptance testing (UAT), is about building stakeholder and user confidence [8].

Testing techniques include functional or "black box" testing and structural or "white box" testing. Functional testing focuses on the external behavior of the software under test. Functional, "black box", testing is based on the requirements or design specification of the software being tested. Structural, "white box", testing focuses on internal structure of the software under test. Test cases designed for structural testing are selected based on the implementation of the software and are designed to execute branches, statements, nodes, or paths [8].

Analysis techniques in testing for determining software quality are either static or dynamic. Static analysis does not include the actual execution of the software being tested. Static analysis techniques are code review, code inspection, and model inspection. Dynamic analysis utilizes real data and real execution, including real

Table 1 Testing methods

Testing method	Scope	Description
Unit	Unit, module, component	Test of basic unit of software
Integration	2 or more software units	Interfaces between components
System	Entire system	End-to-end system testing
Acceptance	Entire system, module	User acceptance and confidence

execution through simulation, of the software being tested. Dynamic analysis techniques can include testing procedures, work instructions, and input of synthetic or real inputs.

2.3 Design and Testing

Designing the tests for a software system is different from testing the design of a software system. Designing test scenarios for a software systems test generally include unit testing, integration testing, system testing, system load testing, and user-defined test cases. Testing the design of a system may include similar tests but generally focuses on usability-related aspects of the system. Designing test scenarios can happen in parallel in both design and development phases.

Testing, development, and design are not necessarily independent activities. The lines between which techniques are specific to design, testing, and development are starting to blur. In the Agile methodology, test-driven development (TDD) incorporates refactoring and test-first development (TFD). Refactoring is making small and incremental change to the design or code without impact on the external behavior of a software system. TFD is the development of a test scenario and writing the code so that the code meets the minimal pass criteria of the test scenario.

2.4 Software Stability

Software evolutionary stability has long been considered an important issue in software engineering research and practice. It is commonly agreed on that software evolutionary stability affects software quality [9]. Specifically, if more frequent and dramatic changes are made to the software product, the likelihood is more errors will be introduced into the code, and accordingly, the original modular design will be compromised and the system will become more difficult to maintain [10]. Therefore, preventing frequent and dramatic changes can lead to higher quality software products. For example, Menzies et al. [11] presented research using the stochastic stability method to avoid drastic software changes.

Component-based software engineering and software product line technology seek to identify and design stable software artifacts, so they can be reused with little or no modifications [12]. For example, Dantas [13] outlined their research plans to define a set of composition-driven metrics to maximize reuse and stability of software modules; Figueiredo et al. [14] studied the evolution of two software product lines, where they analyzed various factors that can affect software stability, including modularity, change propagation, and feature dependency.

Another line of research is to model software stability. Fayad and Altman described a Software Stability Model (SSM) that is used to rearrange relationships between software components in order to accomplish design stability [15, 16].

Their approach is applied to software product lines [17], where the SSM could bring multiple benefits to software architecture, design, and development. The SSM is shown to be significantly better than the existing approaches, but the evaluation criteria are not defined. In other research, Xavier and Naganathan [18] presented a two-dimensional probabilistic model based on random processes to enhance the stability of enterprise computing applications. Their model enables software systems to easily accommodate changes in business policies [19].

Other work in this area includes the identification of stable and unstable software components. Grosser et al. [20] utilized case-based reasoning (CBS) methods to predict stability in object-oriented (OO) software. Bevan and Whitehead [21] used configuration history to locate unstable software components. Wang et al. [22] presented a stability-based component identification method to support coarse-grained component identification and design. Hamza applied the formal concept analysis (FCA) method to identify evolutionary stable components [23].

2.4.1 The Measurement of Stability

There are basically two ways to measure software stability. One way is to analyze a single program and generate its stability metrics based on interdependencies between software modules [24, 25]. These stability metrics reflect the degree of probability that changes made to other modules could require corresponding changes to this module. If a software product has weak dependencies between components, the software is more stable, i.e., change propagation is less likely to happen.

Another way to measure software stability is based on software evolution history, which is why it is also called evolutionary stability. Stability is then represented by differences between two versions of an evolving software product. To measure the differences between two versions of an evolving software product, architecture-level metrics and program-level metrics have been used, which are described below.

Using architecture-level metrics, Nakamura and Basili [26] studied the differences between versions of the same program. In their research, graph kernel theory is used to represent software structure. Comparing the graph difference can provide insight into the version difference. If two systems have different structures, their measurements will always be different. However, architecture-level metrics work best for measuring the stability of an entire software product at the architectural level, but is not readily applicable to measuring the stability of a single component at the source code level.

Using program-level metrics, Kelly [27] introduced the concept of version distance, which measures the differences/similarities between different versions of a software product. In a similar research, Yu and Ramaswamy [28] introduced the concept of structural distance and source code distance. In both of these studies, program-level metrics, such as the number of lines of code, the number of variables, the number of common blocks, and the number of modules, are used as the underlying measures for version distance.

To our knowledge, software evolutionary stability has not been formally measured with information-level metrics based on **Kolmogorov complexity**. This chapter intends to fill this research gap, i.e., using the normalized compression distance (NCD) between two versions of an evolving software product to measure software evolutionary stability.

2.4.2 Kolmogorov Complexity and Normalized Compression Distance

In algorithmic information theory, the Kolmogorov complexity of an object, such as a string (a piece of text), is the length of the string's shortest description in some fixed universal description language [29]. The Kolmogorov complexity of a binary string x can be represented by $K(x)$ denoting the shortest length of string x described using a universal language.

Based on Kolmogorov complexity, Bennett et al. [30] defined information distance between two binary strings x and y as the following.

$$E(x, y) = \max\{(K(x|y), K(y|x)\} \tag{1}$$

In Eq. 1, $K(x|y)$ denotes the conditional Kolmogorov complexity of string x given string y. $E(x, y)$ measures the absolute distance between two strings x and y and thus does not reflect the relative difference between the two strings. Li et al. [31] then defined normalized information distance between two binary strings x and y.

$$d(x, y) = \frac{\max\{(K(x|y), K(y|x)\}}{\max\{K(x), k(y)\}} \tag{2}$$

Equation 2 is based on Kolmogorov complexity, which is non-computable in practice. To make the information distance more practical to use, Li et al. [31] further substituted Kolmogorov complexity $K(x)$ of a string x with the optimal compressed length $C(x)$ of string x. The NCD between two binary strings x and y is then defined [31].

$$\text{NCD}(x, y) = \frac{C(xy) - \min\{(C(x), C(y)\}}{\max\{C(x), C(y)\}} \tag{3}$$

In Eq. 3, xy denotes the concatenation of strings x and y. The NCD has been applied in many fields such as construction of phylogeny trees, music clustering, handwriting recognition, and distinguishing images [32, 33].

Arbuckle et al. [34] are the first to apply NCD in software evolution field. In his series of publications, Arbuckle demonstrated how NCD could be used to visualize software changes [35] and study software evolution [36–38]. Arbuckle's research inspired the work presented in this chapter, i.e., using normalized distance to measure software evolutionary stability.

2.5 Design Structure Matrix

As the size of a project or a process increases, so does the complexity of the system as a whole. When a system reaches a certain size or complexity, it becomes more difficult to see how the system functions as a whole. The DSM was created to specifically address this problem. The decomposition of a large or complex system into less complex, smaller subsystems allows one to break down the problem into smaller and more approachable components. A term coined in the 1970s by Don Steward [39], the DSM's purpose is to give a compact view of an otherwise complicated system.

A DSM will take a complex system and break it down into three basic steps. First, the whole system will be broken down into subsystems of similar parts. Each one of these parts should be easier to address, and each part should almost always function autonomously, not strongly influenced by or heavily reliant on other subsystems. The next step is to pinpoint all of the existing relationships between subsystems and how they interact. This will systematically show how the entire system works as a whole, broken down into steps. The third and final step includes noting all external outputs and inputs used to drive the system. Annotation of how each impacts the system is vital in order to understand how the system works and how it is affected by outside forces.

The DSM could break the entity up even further, taking subsystems and breaking them down into specific elements. From those elements, the system can be broken down further into *system chunks*. These chunks are then assigned to either different individuals or different teams, based on the specificity of the chunk and the skill set of the party assigned to it. Breaking the system down in such a way creates a skeleton and defines its architecture. It also divides the work load. The matrix itself is a square matrix with rows and columns for each of the elements in the system. The following DSM, Fig. 2, illustrates a simplified example.

The matrix shows us the dependencies of each element in the system. Reading the matrix by row reveals each element that the chosen row provides to. For example, element *B* provides to the elements *A*, *C*, *D*, and *F*. Reading the individual columns will give you what each element depends on. Reading the column for element *C*, it is clear that the element depends on *B*, *E*, and *F*.

	A	B	C	D	E	F
A	A					
B	•	B	•	•		•
C	•	•	C		•	
D				D		
E		•	•		E	•
F	•		•	•		F

Fig. 2 DSM simplified example

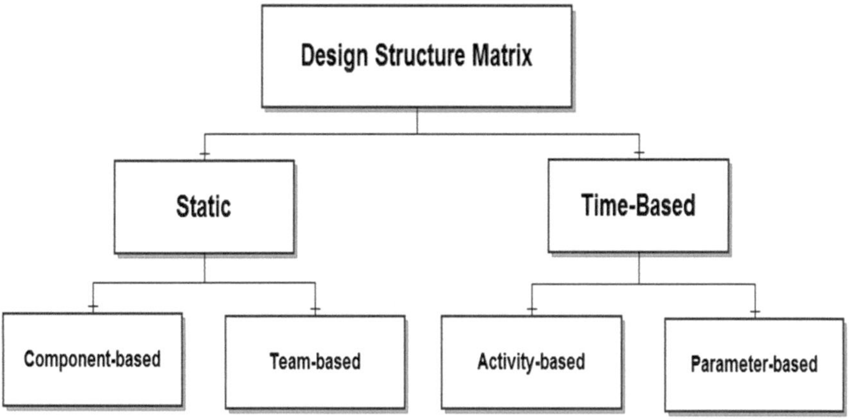

Fig. 3 DSM taxonomy (adapted from [40])

Furthermore, the DSM can be broken up into two different subsets based on linearity. These two different kinds of DSM are known as the static DSM and the time-based DSM. From there, each category can be broken down further into two more applications of DSM. The static DSM can be defined as either component-based (architectural) or team-based (organizational), while the time-based DSM can be broken down into either activity-based (schedule) or parameter-based (low-level schedule). Figure 3 graphically represents the breakdown of the DSM. Each subcategory and application will be explained in further detail in the following sections.

2.5.1 Static DSM

A static DSM represents a system not as it progresses through time but as an unchanging entity. This implies that all of the components of a static DSM coexist. Since no single component receives chronological precedence in the system, these systems can be identified through a process known as clustering algorithms. Next, the static DSM can be branched into two different types as follows.

2.5.2 Component-Based DSM

Component-based DSM is also known as system architecture DSM or product DSM [39]. In a component-based DSM, the entity is broken down into components and/or subsystems of components as well as the specific relationships each component shares. While the architectural decomposition of a system, like any DSM, goes through the three steps, the component-based approach facilitates both systemization and innovation, which cannot be said for every instance of the DSM [40].

Table 2 Quantification scheme for scaling

Required	+2	Necessary for functionality
Desired	+1	Beneficial, but not necessary for functionality
Indifferent	0	No effect on functionality
Undesired	−1	Causes negative effects to functionality but does not prevent it
Detrimental	−2	Must be prevented to achieve functionality

In a component-based DSM, the tallies or marks are often replaced by a negative or positive integer. This integer represents the level of importance of the relationship at hand. Table 2 provides an example of how a quantification system could look. Similar techniques have been successfully used in many approaches that integrate decision-based techniques to quantify decision choices [41]. Figure 4 shows a rudimentary component-based DSM, and Fig. 5 shows an optimized component-based DSM.

By plotting these out on both the columns and rows, we can see the relationship between each individual piece. This may be a rather rudimentary example, but it illustrates how a complex system would look broken down as a component-based DSM in a very simplified way.

	A	B	C	D	E	F
A	A		2			2
B		B				
C	2		C	2		2
D			2	D		
E					E	
F	2		2			F

Fig. 4 Rudimentary component-based DSM

	A	C	F	D	E	B
A	A	2	2			
C	2	B	2			
F	2	2	C	2		
D			2	D		
E					E	
B						F

Fig. 5 Optimized component-based DSM

	A	C	F	D	E	B
A	A	2	2			
C	2	B	2			
F	2	2	C	2		
D			2	D		
E					E	
B						F

Fig. 6 Team-based DSM

2.5.3 Team-Based DSM

Instead of components, the team-based DSM is organized based on the people or teams and their interaction. This is very similar to the component-based approach with the added layer of teams. In essence, a team-based DSM can be looked at as a component-based DSM with an extra assignment layer where responsibilities are distributed based on the skill set and ability of each team.

Taking sections of a component-based DSM and highlighting the different areas of responsibility give insight as to how the workload is divided. Overlapping areas of the workload show where multiple teams may have to coexist in regard to responsibility. These areas should be chosen based on the dependency of components. Thus, the ordering of component along the axis should be based on the groupings of team responsibility with highly dependent components toward the edge of each team's area of responsibility, similar to a Venn diagram.

In Fig. 6, this is illustrated. Two different teams are used to split the responsibility of the elements while both teams share the responsibility of elements *F* and *C*, which are dependent of each other.

2.5.4 Time-Based DSM

In a time-based DSM, the ordering of the rows and columns coincides with the flow of time. Activities are marked as either upstream or downstream.

2.5.5 Activity-Based

This type of DSM is appropriate when a system is built on a group of activities with a specific schedule. The matrix is then designed chronologically by order of the activity. This makes more sense with a time-sensitive project or process and is why this DSM is also referred to as process based.

One can visualize what the DSM would look like based on the previous exemplar DSMs. The difference lies in the elements. Now elements are replaced with time-sensitive processes that take chronological precedence from one another. In that sense, the processes are automatically grouped by dependence since a process depends most likely on the feedback, or the previous process. The goal of creating the DSM and analyzing the process is typically to reduce duration.

2.5.6 Parameter-Based

When the modeling of a system is based on low-level relationships between design decisions and parameters; this (parameter-based DSM) is used [42]. This is, in some ways, a further breakdown of the activity-based DSM. The difference between activity-based and parameter-based analysis is another dimension. Not only does the parameter-based DSM take chronology into consideration, it also accounts for multiple variable differences. Parameter-based analysis factors in different inputs to compare various outcomes.

Overall, each type of DSM produces similar results. It is a way to simplify otherwise extremely intricate systems or processes. Problems such as dependencies, time constraints, and variables can be observed through a dissection of systems in order to plan issue and risk mitigation. Table 3 illustrates the different types of DSM.

Table 3 DSM list and usages [42]

DSM Type	Representation	Applications	Integration analysis
Component-based	Components in a product architecture and their relationships	System architecting, engineering, design.	Clustering
Team-based	Individuals, groups, teams and their relationships	Organization design, interface management	Clustering
Activity-based	Activities in a process and inputs/outputs	Project scheduling, activity sequencing, cycle time reduction, risk reduction	Sequencing
Parameter-based	Design parameters and their relationships	Low-level process sequencing and integration	Sequencing

3 Evolutionary Stability and the mDSM Methodology

In this section, we introduce evolutionary stability metrics for software systems and the mDSM Methodology.

3.1 *Software Evolutionary Stability Metrics*

To discuss evolutionary stability, it is necessary to measure the difference between two versions of an evolving software artifact. As described before, program-level measures have been used to measure the *distance* between evolving software components [27, 8]. In this chapter, NCD is used to study the difference between two versions of a software artifact.

(Definition)Version distance: Let m and n be the two versions of a software artifact. The *version distance* between m and n is defined to be $\mathrm{VD}(m,n) = \mathrm{NCD}(S_m, S_n)$, where S_m and S_n are the binary string representations of the source code of m and n, respectively.

In the definition of version distance, m and n can be single files, which will be easily converted to single strings, or they can contain multiple files, which can be merged (concatenated) as single strings. It should be noted that although the formal definition of *version distance* is new in this paper, it has been implicitly used in Arbuckle's studies [33–37]. Version distance VD(m, n) between versions m and n could fall in the range of [0, 1], where 0 means the two versions are exactly same and 1 indicates the two versions have the maximum possible difference.

Considering the evolution of a software program shown in Fig. 7 (each box represents one version/release—arrows represent the dependencies between versions and each row represents one branch), it is not a trivial task to measure the evolutionary stability of this product. On one hand, this program contains many branches and many versions. On the other hand, software stability is an empirical observation; it is directly dependent on the length of the observation period. To simplify the problem, we introduce several metrics to measure the evolutionary stability of a product with respect to different concerns instead of presenting a single evolutionary stability metric for the whole product.

(Definition) Version stability: Given one version of a software artifact (say m), its version stability $\mathrm{VS}(m)$ is defined as 1 minus the average version distance between version m and other subsequent p versions (n_1, n_2, ..., n_p), which are directly or indirectly evolved from m and released in a given period of time.

$$\mathrm{VS}(m) = 1 - \sum_{i=1}^{p} VD(m, n_i)/p \tag{4}$$

If we consider the evolution of a software product as a tree structure, in the definition of version stability, version m will be considered as the ancestor of its

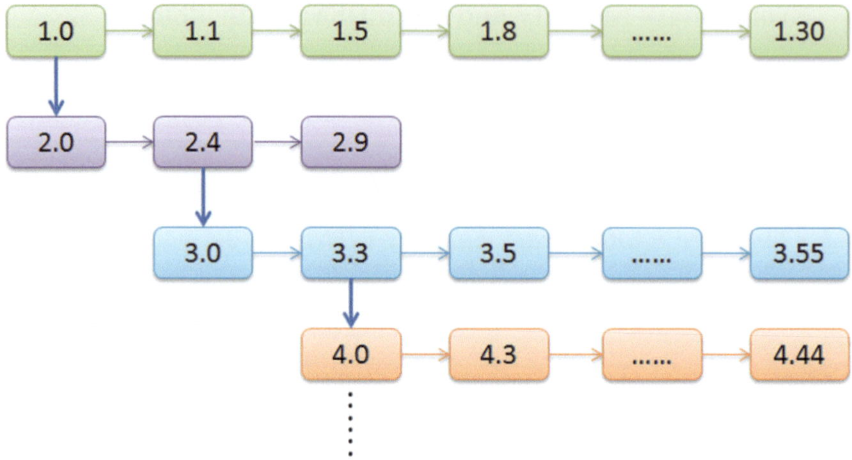

Fig. 7 The evolution of a software program

subsequent versions. Consider the stability of Version 1.1 in Fig. 7. Assume within one year after the release of Version 1.1, two subsequent versions based on 1.1 are released: 1.5 and 1.8. The stability of Version 1.1 in this period can be calculated as

$$\mathrm{VS}(1.1) = 1 - (\mathrm{VD}(1.1, 1.5) + \mathrm{VD}(1.1, 1.8))/2.$$

The concept of version distance has a reverse relation with the general concept of software stability, where long distance means low stability and short distance means high stability. This relation is reflected in Eq. 4. The value of version stability could then fall in the range of [0, 1], where 0 means the lowest stability and 1 means the highest stability.

(**Definition**) **Inter-Component Version stability**: Given one version of a software artifact may have multiple interacting software components (say c), the inter-component version stability $\mathrm{ICVS}(c)$ is defined as the average version stability among an interacting software artifacts' components m_{ci} and other subsequent p components $(c1, c2, \ldots, cp)$.

$$\mathbf{ICVS}(\boldsymbol{c}) = \sum_{i=1}^{p} \mathbf{VS}(\boldsymbol{m}_{ci})/\boldsymbol{p} \tag{5}$$

The value of inter-component version stability is in the range of [0, 1], where 0 means the lowest stability and 1 means the highest stability among interacting components.

(**Definition**) **Branch stability**: Given one branch (say b) of a software artifact that has released p versions in a given period of time, its branch stability $\mathrm{BS}(b)$ in this period is defined as 1 minus the average version distance between every two releases in p.

$$\mathrm{BS}(b) = 1 - \sum_{m \in p} \sum_{n \in p, n \neq m} \mathrm{VD}(m,n)/C_p^2 \tag{6}$$

Consider the stability of Branch 2.0 in Fig. 7, which contains 3 versions (releases). According to Eq. 6, its stability can be calculated as $\mathrm{BS}(2.0) = 1 - (\mathrm{VD}(2.0, 2.4) + \mathrm{VD}(2.0, 2.9) + \mathrm{VD}(2.4, 2.9))/3$. In our definition of branch stability, we measure the distance between every two releases instead of the distance between two consecutive releases. Our definition captures the evolution of the entire branch as a whole. The version distance between every pair of consecutive releases can be small, but the earlier release and later release can be dramatically different if every change is performed on different parts of the artifact.

(**Definition**) **Structure stability**: Given an artifact (say a) of a software product that has p major releases in a given period of time, its structure stability $\mathrm{SS}(a)$ in this period is defined as 1 minus the average version distance between every two major releases in p.

$$\mathrm{SS}(a) = 1 - \sum_{m \in p} \sum_{n \in p, n \neq m} \mathrm{VD}(m,n)/C_p^2 \tag{7}$$

In the definition of structure stability, major releases refer to those versions that experienced dramatic changes comparing with their previous versions. For example in Fig. 7, Versions 1.0, 2.0, 3.0, and 4.0 can be considered as major releases in a specified period. The assumption behind this definition is that major releases most likely will include some structure changes.

(**Definition**) **Aggregate stability**: Given an artifact (say a) of a software product that has released p versions in a given period of time, its aggregate stability $\mathrm{AS}(a)$ in this period is defined as 1 minus the average version distance between every two releases in p.

$$\mathrm{AS}(a) = 1 - \sum_{m \in p} \sum_{n \in p, n \neq m} \mathrm{VD}(m,n)/C_p^2 \tag{8}$$

Aggregate stability measures the entire evolution of the artifact. The computing process might be expensive and time-consuming. It might be a feasible metric for a product with small number of releases, however, if a system contains too many versions, such as Linux with over 700 releases, structure stability is a better substitute for aggregate stability.

To make these definitions more understandable, we make the following remarks. First, the above definitions are applicable to any software artifacts that can be documented as pieces of text or strings, such as source code, requirement specification, testing results, and design specification. Second, there is no limitation on the size of the software artifact. It can be a single component, across multiple components, a file or the whole system, as long as it is computationally feasible.

3.2 mDSM Methodology

A component-based DSM is extended to realize the metrics-based DSM (mDSM). The mDSM methodology follows: (i) decomposition of the system into elements, subsystems, or components, (ii) documentation and understanding the interaction and integration of the elements, expressed as semantic rules, (iii) calculate the evolutionary stability of the component(s), (iv) lay out the square mDSM with components labeling rows and columns, grouped by subsystems or modules and represent the interactions using evolutionary stability values of selected metric in the mDSM cells, displaying the DSM, and (v) analysis of potential reintegration points clustering or integration analysis [39, 40]. In the process of decomposing the elements (modules or subsystems), the elements will also be defined. The definitions are presented to clarify their use within the system and to promote greater understanding of complex systems relationships. Defining elements will also assist in the documentation and understanding of the element interactions.

3.2.1 Decomposition and Modularization of the Software System

Decomposition of the software system is simply the breaking down or factoring the software system into smaller and more manageable elements, subsystems, units, modules, or components. This is the first step in the mDSM process and will be clearly shown in Sect. 4. Modularization is a beneficial technique for reducing interdependencies among components of system [43]. Removing unnecessary dependencies and systems redundancy is critical to an efficiently running system. The mDSM can reduce unnecessary dependencies through effective cluster analysis, by the reordering of rows and columns in the matrix [40].

3.2.2 Decompose the Software System into Semantic Rules

Step 2 in the development of the mDSM is to identify and document the interaction and integration of the subsystems or components. Understanding the direct interactions, or dependencies, among systems elements or subsystems is essential to developing a complete mDSM. There are two distinct types of direct data flow interactions, dependencies, in mDSM: inputs and outputs. Inputs are traditionally annotated by an “X” in a standard DSM at the row and column input pairing at the end of each rule. Outputs are traditionally annotated by an “X” in a standard DSM at the column and row output pairing at the end of each rule. Defining inputs and outputs allows for the establishment of semantic rules to describe the systems interactions to be mapped within the mDSM.

3.2.3 Calculate the Evolutionary Stability Metric

We recommend using an evolutionary stability metric suitable to the mDSM that evaluates the dependencies, or the interactions across the components. As version stability is extended to inter-component version stability for this purpose of the mDSM: the same could be accomplished with branch stability, structural stability, and aggregate stability. The mDSM in this chapter will utilize inter-component version stability as the metric.

3.2.4 Display the mDSM

The mDSM is built off the decomposition of the system. The mDSM is populated by utilizing the semantic rules that define the systems interactions, established in Sect. 4.2. The original DSM convention is In Rows (IR) notation which places inputs in rows and outputs in columns [39]. The same IR convention is used for the mDSM, yet in the place of a traditional “X” notation or simple quantification scheme, a value for the selected evolutionary stability metric will be used.

3.2.5 Clustering

Clustering can be performed via distance penalty algorithms [44] or inspection of any mDSM. Often, software systems are already componentized or modularized. Clustering can provide opportunities for improvement and optimization in testing and design.

3.2.6 mDSM Utilizing Evolutionary Stability Metrics

The mDSM utilizing evolutionary stability metrics will render as a matrix containing a value between [0, 1]. 0 pertains to low evolutionary stability and 1 is high evolutionary stability. Figure 8 shows a simple mDSM, where A through F are a systems module and the values, across direct information-level interactions, are the inter-component version stability of the interactions. Each module is annotated on the row and column levels.

Fig. 8 Example mDSM

	A	B	C	D	E	F
A	A					
B		B				
C		.78	C			
D	.91			D	.55	
E					E	
F						F

4 Case Studies

The Apache Ant and Apache HTTP studies are to introduce a method of measurement of software evolutionary stability utilizing information-level metrics based upon Kolmogorov complexity. These open-source systems case studies in Sects. 4.1.1 and 4.1.2 to establish the feasibility and methodology of computation are critical to the refinement of the mDSM. The closed-source system case study in Sect. 4.2 builds upon the use of the evolutionary stability metrics applying to a much more complex system. The mDSM provides a holistic view of a complex systems evolutionary stability.

4.1 Apache Ant and HTTP with Evolutionary Stability Metrics

In this chapter, we measure the evolutionary stability of two open-source products, Apache Ant and Apache HTTP. All available source files were downloaded from the Apache Software Foundation [38]. The general information about these two products is summarized in Table 4. For each release (version), all the source code files are concatenated to create a single text file (string). The program used to measure the NCD between two files is implemented using C++ with the 7-Zip compressor [39] in the Windows environment.

In the remainder of this section, we report the results of our studies on various metrics of software evolutionary stability of Apache Ant and Apache HTTP.

Table 4 Summary of Apache Ant and Apache HTTP (up to September 2011)

	Ant	HTTP
Category	Build automation tool	Web server
Language	Java	C
Source code file	*.java	*.c, *.h
Initial release	2000	1995
Number of branches	8	5
Total releases	21	84

4.1.1 Apache Ant

Figure 9 illustrates the evolution tree of Apache Ant. It is worth noting that in Branches 1.1, 1.2, and 1.3, there is only one release in each of them.

To study version stability according to our definition, there must be some subsequent releases based on this version. In the case of Apache Ant, we used the time frame of two subsequent releases to calculate the version stability. Eight out of the 21 releases satisfy the time frame requirement. Their version stability is shown in Table 5. It can be seen that in our study period, the most stable version is 1.6.3,

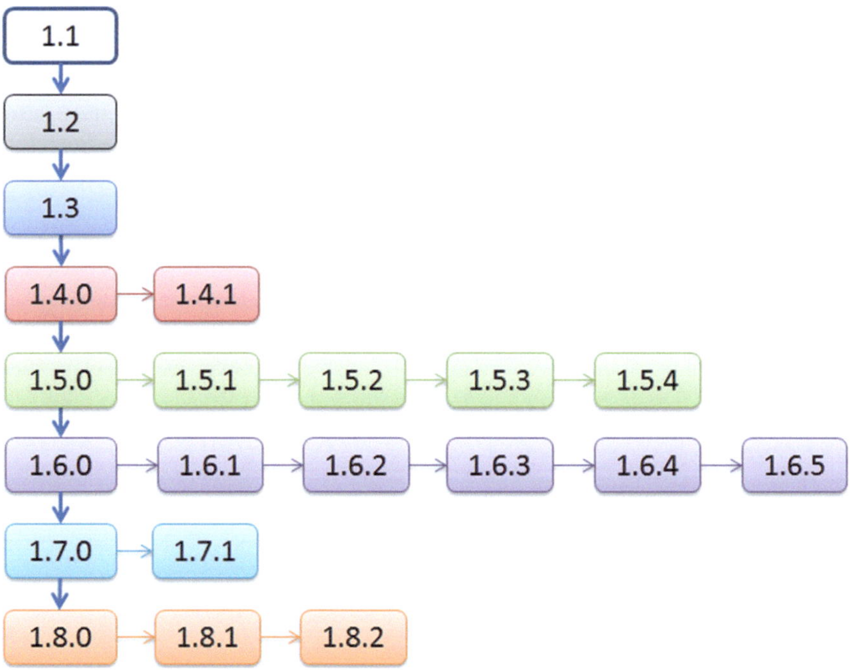

Fig. 9 The evolution tree of Apache Ant

Table 5 Version stability of selected Ant releases

Version	1.5.0	1.5.1	1.5.2	1.6.0	1.6.1	1.6.2	1.6.3	1.8.0
Subsequent releases used in the measurement	1.5.1 1.5.2	1.5.2 1.5.3	1.5.3 1.5.4	1.6.1 1.6.2	1.6.2 1.6.3	1.6.3 1.6.4	1.6.4 1.6.5	1.8.1 1.8.2
Version stability	0.956	0.933	0.994	0.932	0.914	0.924	0.998	0.976

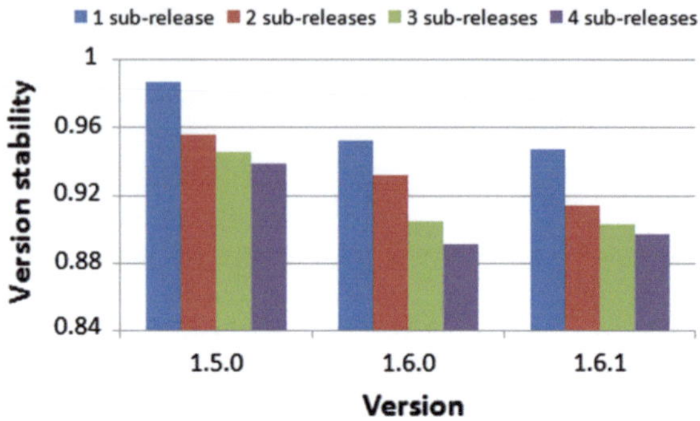

Fig. 10 Version stability calculated with different evolution time frames

which has version stability 0.998; the most unstable version is 1.6.1, which has version stability 0.914. These values can be interpreted as the following: From Version 1.6.3 to Versions 1.6.4 and 1.6.5, 99.8 % of the source code remains unchanged; From Version 1.6.1 to Versions 1.6.2 and 1.6.3, 91.4 % of the source code remains unchanged.

The measurement of version stability is expected to decrease with the increase of evolution time frame. This observation is illustrated in Fig. 10, where the evolution stabilities of Versions 1.5.0, 1.6.0, and 1.6.1 are measured with different time frames. Figure 10 further indicates that software evolution stability is a time-dependent concept. We can only measure a software product's evolutionary stability within a certain time period. There is no absolute measurement of a product's evolutionary stability.

From Fig. 9, we can see that the two longest evolved branches in Ant are 1.5 and 1.6. To see which of these two branches is relatively stable, we calculate their branch stability. The result shown in Table 6 illustrates that Branch 1.5 is relatively more stable than Branch 1.6. The result is the same if we compare all five releases

Table 6 Branch stability of Ant 1.5 and 1.6

Branch	1.5	1.6	1.6
Versions used in the calculation	All 5 releases	First 5 releases	All 6 releases
Branch stability	0.953	0.912	0.918

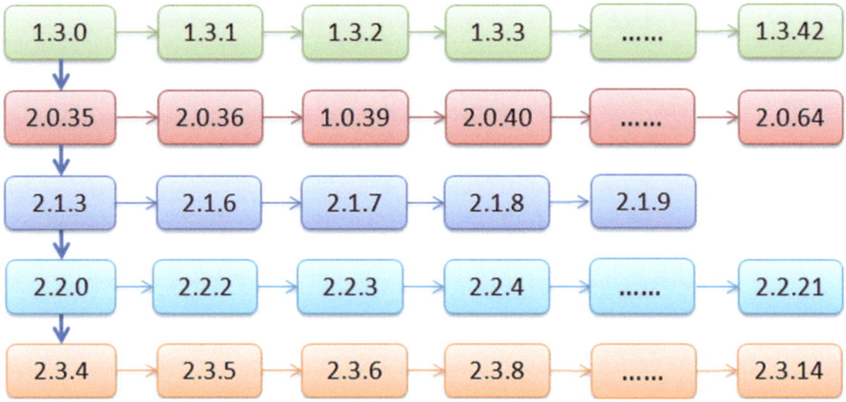

Fig. 11 The evolution tree of Apache HTTP

of Branch 1.5 with (a) the first five releases of Branch 1.6 and with (b) the total six releases of Branch 1.6. Our definition of branch stability (Eq. 6) is based on the measurement of version distance of every pair of releases in each branch. Therefore, it reflects the stability of the entire branch as a whole.

4.1.2 Apache HTTP

Figure 11 illustrates the evolution tree of Apache HTTP. There are five branches and total 84 releases.

Because Branches 1.3, 2.0, and 2.2 have more releases than other branches (2.1 and 2.3), they are selected to study the general trend of version distance. Figure 12 shows the version distance between the first release and its subsequent releases, and the last release and its previous releases in Branches 1.3, 2.0, and 2.2. It can be seen that major changes appear in the earlier releases while minor changes appear in the latter releases, which are reflected by the growth trend of version distances, i.e., the version distance becomes more stable with the increasing of time.

To further verify our observations, version stability of selected releases is studied and the result is shown in Fig. 13. The version stability of these releases is calculated using the subsequent 10-release time frame. It can be seen in all three branches, with the selected versions, that with revision versions become more stable over time.

Figure 14 illustrates the branch stability of the five branches of Apache HTTP. Two measures of branch stability are calculated: (1) using all releases in each branch and (2) using the first five releases in each branch. The latter case is in accordance with the total number releases in Branch 2.1. It can be seen

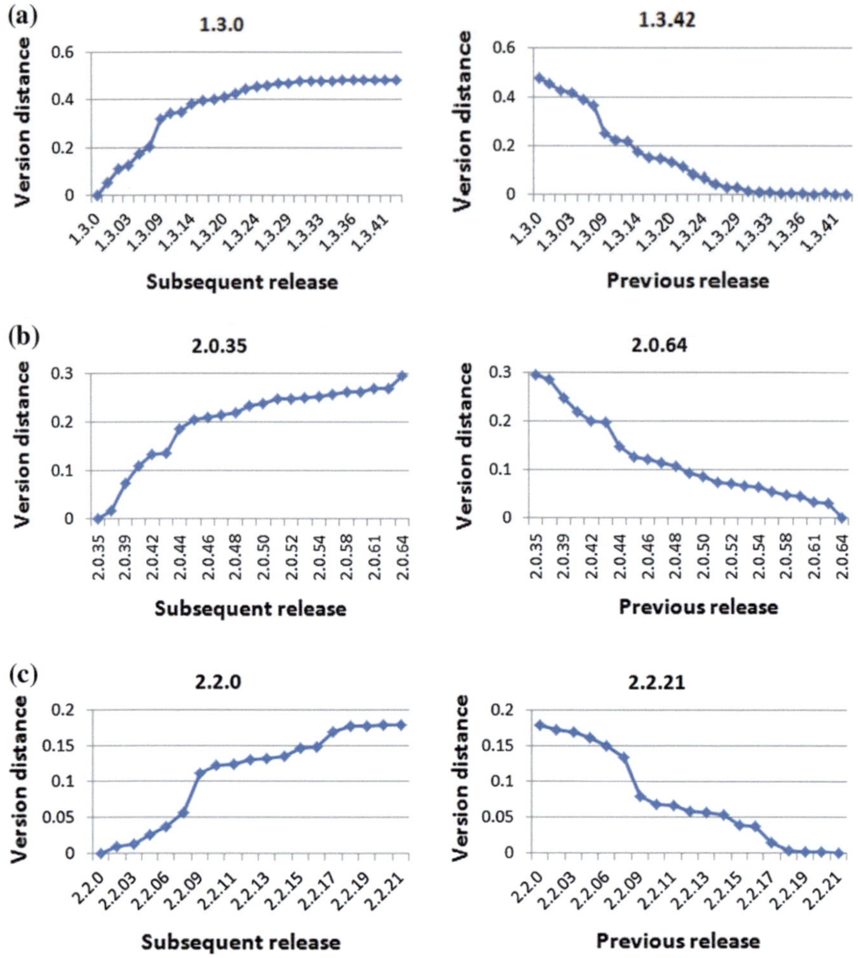

Fig. 12 Version distance between the first release and its subsequent releases and the last release and its previous releases in Branches **a** 1.3; **b** 2.0; and **c** 2.2

that Branch 2.2 is most stable in the first five releases while Branch 2.1 is most stable for all the releases. Branch 1.3 is most unstable in both the two measurements.

4.1.3 Comparing Apache Ant and Apache HTTP

We studied the version stability and branch stability of Apache Ant and Apache HTTP in the previous two subsections. In this subsection, we compare the structure stability and aggregate stability of these two products.

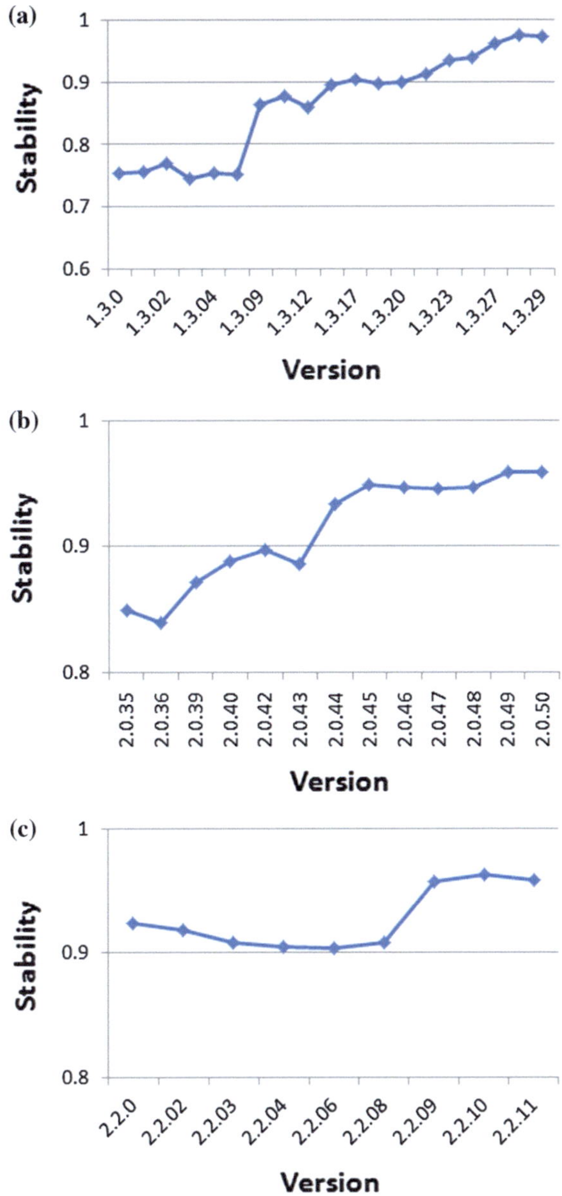

Fig. 13 Version stability of selected releases in Branches **a** 1.3; **b** 2.0; and **c** 2.2

Figures 15 and 16 show the heat map of version distance between major releases of Apache Ant and Apache HTTP, respectively. Some long version distances are found between earlier releases and later releases in Apache Ant. Table 7 further shows that Apache HTTP is structurally more stable than Apache Ant.

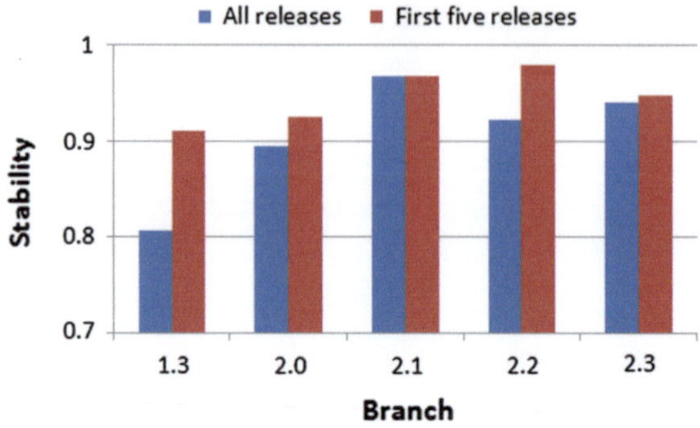

Fig. 14 Branch stability of Apache HTTP

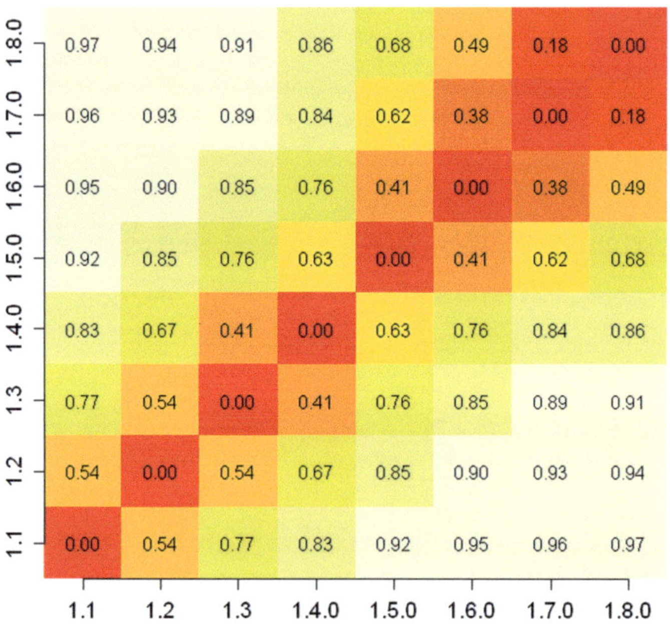

Fig. 15 Heat map of version distances between major releases of Apache Ant

The aggregate stability in Table 7 also shows that Apache HTTP is a bit more stable than Apache Ant despite the fact that Apache HTTP has more releases than Apache Ant.

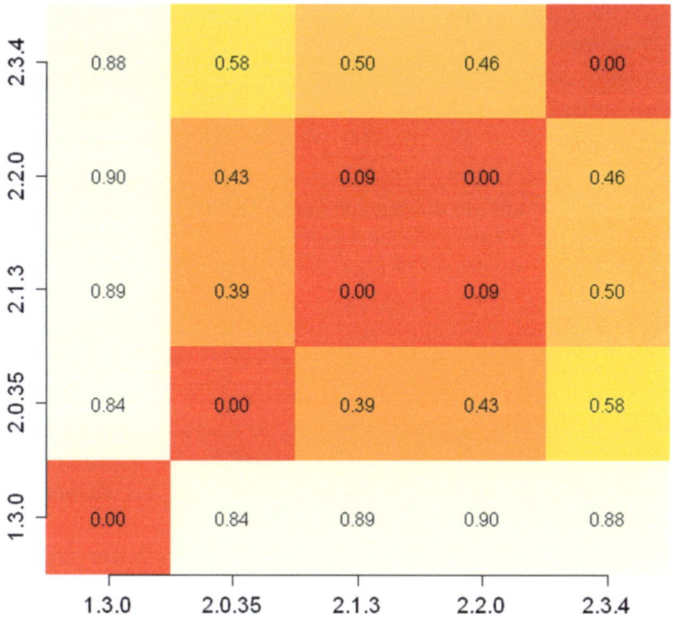

Fig. 16 Heat map of version distance between major releases of Apache HTTP

Table 7 Structure stability and aggregate stability of Apache Ant and Apache HTTP [46, 47]

	Ant	HTTP
Number of major releases	8	5
Number of releases	21	84
Structure stability	0.270	0.403
Aggregate stability	0.441	0.470

4.2 Enterprise Study with mDSM

We performed a case study on the enterprise resource planning (ERP) system of a multi-national conglomerate with diverse product lines in defense, aerospace, and commercial with a combined total of approximately 40,000 employees and 12 billion dollars in annual net revenue. The organization has unsuccessfully attempted to replace their ERP, in whole, once every five years or in parts (replacement of a sub-system), every two years for the last 10 years. While multiple factors play into these system implementation failures, common factors exist: the lack of stakeholder understanding of systems interactions, how those interactions impact other systems in the system of systems, and by how the systems stakeholders define successful outcomes across systems metrics. Regardless of the number of consultants and toolsets used, the stakeholders did not accurately assess the complexities of their systems. A new tool is needed to rationalize and aide future architectural decisions.

This particular complex software system was chosen for case study for three reasons: The system meets the definition of a complex system, the number of system implementation failures, and the author's ability to study and access the system of systems. The systems interactions of interest for this study are data flow interactions. The data flow interactions are of particular interest because the organization is managed via these data flow interactions. A system within the system either inputs data to a subsystem or outputs data to a subsystem.

The ERP system of this organization is modular: loosely coupled between modules and highly cohesive within modules. The technological landscape is very diverse—it includes DB2 on IBM i-Series (AS400), Microsoft SQL Server on Microsoft Windows, and Oracle on Unix. Users interact with the data through Java and .Net web applications, dumb terminals (green screen), and various reporting technologies. While this case study's subject is ERP, the same process is applicable to any system: natural or artificial. The metrics chosen and the interactions explored would merely be defined to the systems requirements.

ERP systems are complex software systems that support organizations. ERP systems support organizations functional areas such as operations, engineering, finance, human resources, and trade compliance. ERP system implementations are difficult, costly, and prone to failure. High-profile ERP implementation failures are well documented across multiple industries [1, 4]. This highlights the need for the rationalization of software systems during design and testing.

4.2.1 Case ERP Decomposition

The case study ERP is made up of operations, engineering, trade compliance, finance, human resources (HR), and customer relationship management (CRM) systems. **(i) Operations**. The operations, or supply chain, system can be decomposed into the following systems: (a) capacity planning—utilization of machines, people, and bottleneck avoidance, (b) inventory management—managing raw materials through finished goods, (c) manufacturing execution systems (MES)—tracking materials and labor for overall cost of production, work instructions, machine interactions (programmable logic controllers, PLCs), and quality tracking, (d) master planning/forecasting—business development forecast allows planning for resources, materials, timing, and planned orders, (e) materials resource planning (MRP)—generates all material needs in the form of requisitions or purchase orders for actual customer orders (firm) and planned orders, (f) order processing—all actual customer (firm) orders, (g) procurement—purchasing materials or services, (h) production scheduling—daily production schedule for the manufacturing floor, (i) receiving—receipt of shipments, (j) shipping—final inspection and generation of invoice, shipping of product. **(ii) Engineering**. The engineering system is decomposed into the following systems: (a) computer-aided design (CAD)—design and product drawings, (b) revision control—changes to any specification in product or projects, (c) product life cycle management (PLM)—life cycle of the product from order to ship, engineering drawings, and specifications. **(iii) Trade Compliance**. The trade compliance system has no sub-systems. Trade compliance is responsible for regulation of vendors and vetting of customers prior to shipment. **(iv) Finance**. The finance system can be decomposed into the following systems: (a) accounts payable (AP)—tracks and pays invoices, (b) accounts receivable (AR)—cash collection against shipping invoices, (c) earned value management (EVM)—calculates the progress of a product or service against a project plan, (d) fixed assets (FA)—tracks depreciable assets, (e) general ledger (GL)—tracks all financial transactions, (f) payroll—payment for labor, (g) project/cost accounting—tracks actual costs to a project, important in cost plus contracts, (h) timekeeping—labor tracking. **(iv) Human Resources (HR)**. The human resources system is made up of the following systems: (a) benefits—management of benefit providers and employee benefits, (b) learning management system (LMS), (c) skills matrix—skill sets of employees across the organization, (d) succession planning—development and plan for future organizational leadership. **(v) Customer Relationship Management (CRM)**. The customer relationship management system is utilized by business development to manage sales and customer service. This system is not broken down into any sub-systems. The semantic rules in Table 8 are broken down by row, column pairings within the mDSM matrix. For example, semantic rule 1, capacity planning inputs into FA (r:1, c:18), is annotated in row 1, column 18 of the mDSM matrix.

4.2.2 Calculate the Inter-component Version Stability

Figure 17 shows the major revisions evolution tree of the case ERP over the last 10 years. In order to extract and infer additional information from each component interaction, we calculate the inter-component version stability (Eq. 5) of each inter-component interaction. As in Sect. 4.1, we again utilize a program to measure the NCD between two component files as implemented using C++ with the 7-Zip compressor [39] in the Windows environment. For example, using semantic rule number 1, we calculate the version stability of row 1, column 18, capacity planning (CP) and the version stability of FA, using Eq. 4. The results of VS of CP and VS of FA are then calculated with Eq. 5. Therefore, using Eq. 4, VS of CP (11.0.2.0) = 1 − (VD(11.0.2.0, 11.0.3.2) + VD(11.0.2.0, 11.1.1.2))/2 and VS of FA(11.0.2.0) = 1 − (VD(11.0.2.0, 11.0.3.2) + VD(11.0.2.0, 11.1.1.2))/2. Using Eq. 5, VS of CP(11.0.2.0) + VS of FA(11.0.2.0)/2 gives us the inter-component version stability of capacity planning and fixed assets components. The result is a number between [0, 1] with 0 exhibiting the lowest stability and 1 being stable. This is calculated for every interaction within the mDSM.

Table 8 Semantic rules

1. Capacity Planning inputs into Fixed Assets (r:1, c:18)	25. Shipping outputs to Order Processing (c:10, r:6)
2. Capacity Planning inputs into Time Keeping (r:1, c:19)	26. Order Processing inputs into Shipping (r:6, c:10)
3. Manufacturing Execution Systems inputs into Order Processing (r:3, c:6)	27. Order Processing outputs to Shipping (c:6, r:10)
4. Manufacturing Execution Systems inputs into Inventory Mgmt (r:3, c:2)	28. Inventory management inputs into Shipping (r:2, c:10)
5. Master Planning/Forecasting inputs into Order Processing (r:4, c:6)	29. Inventory Management outputs to Shipping (c:2, r:10)
6. Master Planning/Forecasting inputs into Capacity Planning (r:4, c:1)	30. Inventory Management outputs to GL (c:2, r:19)
7. Master Planning/Forecasting inputs into Inventory Management (r:4, c:2)	31. Receiving inputs into Inventory Management (r:9, c:2)
8. Material Resource Planning inputs into Inventory Management (r:5, c:1)	32. Receiving inputs into Procurement (r:9, c:7)
9. Material Resource Planning inputs into Order Processing (r:5, c:6)	33. Receiving outputs to Inventory Management (c:9, r:2)
10. Material Resource Planning inputs into Procurement (r:5, c:7)	34. Receiving outputs to Accounts Payable (c:9, r:16)
11. Order Processing inputs into Capacity Planning (r:6, c:1)	35. Accounts Payable inputs into GL (r:15, c:19)
12. Order Processing inputs into Trade Compliance (r:6, c:14)	36. Accounts Payable outputs to GL (c:15, r:19)
13. Procurement inputs into Material Resource Planning (r:7, c:5)	37. Accounts Receivable inputs into GL (r:16, c:19)
14. Procurement inputs into Trade Compliance (r:7, c:14)	38. Accounts Receivable outputs to GL (c:16, r:19)
15. Procurement outputs to GL (c:7, r:19)	39. Benefits outputs to Payroll (c:23, r:20)
16. Production Scheduling inputs into Capacity Planning (r:8, c:1)	40. Timekeeping outputs to Payroll (c:22, r:20)
17. Production Scheduling inputs into Order Processing (r:8, c:6)	41. Payroll inputs into Time Keeping (r:20, c:22)
18. Production Scheduling inputs into Inventory Management (r:8, c:2)	42. Payroll outputs to GL (c:20, r:19)
19. Production Scheduling inputs into Timekeeping (r:8, c:22)	43. Computer Aided Design inputs into PLM (r:11, c:12)
20. Shipping inputs into Inventory Management (r:10, c:2)	44. Computer Aided Design outputs to PLM (c:11, r:12)
21. Shipping inputs into Order Processing (r:10, c:6)	45. Revision Control inputs into PLM (r:13, c:12)
22. Shipping inputs into Trade Compliance (r:10, c:6)	46. Revision Control outputs to PLM (c:13, r:12)

(continued)

Table 8 (continued)

23. Shipping outputs to Inventory Management (c:10, r:2)	47. Skills Matrix inputs into Learning Management System (r:25, c:24)
24. Shipping outputs to Account Receivables (c:10, r:16)	48. Succession Planning inputs into Skills Matrix (r:26, c:25)

Fig. 17 The evolution tree of case ERP

			1	2	3	4	5	6	7	8	9	10	11	12	13	14	15	16	17	18	19	20	21	22	23	24	25	26	27
Operations	Capacity Planning	1	1																	0.89	0.91								
	Inventory Management	2		2							0.85	0.93																	
	MES	3		0.93	3			0.80																					
	Master Planning/Forecasting	4	0.89	0.91		4		0.86																					
	MRP	5	0.87				5	0.82	0.84																				
	Order Processing	6	0.78					6				0.86				0.25													
	Procurement	7					0.92		7							0.20													
	Production Scheduling	8	0.86	0.91				0.87		8														0.45					
	Receiving	9		0.85					0.85		9																		
	Shipping	10		0.93				0.86				10				0.11													
Engineering	Computer Aided Design	11											11	0.45															
	Product Lifecycle Management	12											0.45	12	0.65														
	Revision Control	13												0.65	13														
Trade Compliance	Trade Compliance	14														14													
Finance	Accounts Payable	15									0.90						15				0.87								
	Accounts Receivable	16										0.86						16			0.81								
	Earned Value Management	17																	17										
	Fixed Assets	18																		18									
	General Ledger	19		0.91					0.87								0.87	0.81			19	0.65							
	Payroll	20																				20		0.47	0.30				
	Project/Cost Accounting	21																					21						
	Time Keeping	22																						22					
Human Resources	Benefits	23																							23				
	Learning Management System	24																								24			
	Skills Matrix	25																								0.95	25		
	Succession Planning	26																									0.98	26	
CRM	CRM	27																											27

Fig. 18 mDSM of case ERP

4.2.3 Render the Case ERP mDSM

The Case ERP mDSM is rendered by placing the calculated inter-component evolutionary stability value on every direct dependency within the software system. Once rendered, we can inspect the stability of the interactions of the software system. The case ERP mDSM is shown in Fig. 18.

4.3 Results of mDSM Using Evolutionary Stability Metric

Overall, the case ERP is very stable as realized in the mDSM in Fig. 18. The use of the evolutionary stability metric, inter-component version stability, in the mDSM provides information for where we should focus design and testing efforts. The evolutionary stability of components within the operations module is very high. This is the result of manufacturing processes that have not changed in nearly 20 years. The case ERPs high level of stability may explain the difficulty of replacing the overall ERP. Following, we will examine the less stable interactions.

Examining column 14, showing the interaction of the trade compliance components interaction with order processing at row 6, procurement at row 7, and shipping at row 10, exhibits low evolutionary stability over the last 10 years of the systems evolution. The trade compliance component interaction with above-stated components is where systems design and testing efforts should be focused. Compliance in trade had become a very important issue for the organization, and efforts were made to address new trade regulations, automate compliance, and to avoid fines from the US government. That led to multiple changes in the trade compliance component and to instability with highly evolved system components in the operations module.

In rows 11 and 12, CAD and PLM, respectively, we find a lower than average evolutionary stability. Again, a place where design and testing should focus. As CAD programs were upgraded over the last 10 years, the requirements of the data captured from the CAD drawings were expanded and required structural changes at both the application and data interaction level.

In row 20, the Payroll component, we see lower than average evolutionary stability with the timekeeping and benefits components. Nearly continuous design, development, and testing efforts take place to manage the interaction among these components. Yearly changes to benefits providers and worldwide tax law changes contribute to the lack of stability in these components.

4.4 Impacts to Design and Testing

4.4.1 Abstraction

Abstraction in software systems engineering is the technique of removing the complexity from the intricate details out of the design or testing of a system. Software system testers, designers, and stakeholders have very little need to understand the system at the level of detail of a systems developer or even systems architect. The mDSM utilizing evolutionary stability allows for abstraction: it provides a high-level view of the system that allows understanding and rationalization of the software system as whole without knowing the lower level details.

Abstraction also provides a clear view of potential integration and interface issues that can drive testing and design decisions.

4.4.2 Traceability

Traceability refers to how changes to one artifact impact other artifacts and therefore set in motion a cascade of changes [43]. Understanding downstream impact is important to future state systems and change management, especially when understanding impact of systems replacement. The mDSM provides developers, testers, designers, and stakeholders with the knowledge of how systems control will be impacted and upfront knowledge to potential downstream system impacts.

4.4.3 Optimization

The mDSM provides a matrix view that can be optimized through clustering. The case ERP is pre-clustered by its higher level modules. Optimization of effort for testing and design are keys in this case. The mDSM utilizing an evolutionary stability metric provides clarity to where development, testing, and design efforts must be focused.

4.4.4 Boundaries

Viewing the lower triangular of the mDSM, we expect to see higher values for evolutionary stability. On the lower triangular, we expect to see generally lower values of evolutionary stability. The expectation for this in the case ERP is that the system in its entirety is highly evolutionarily stable and very modular. We expect *high cohesion* within in the higher level modules of operations, engineering, trade compliance, finance, human resources, and CRM. And *loose coupling* between the higher level modules. High stability in the highly cohesive lower triangular and lower stability in the loosely coupled upper triangular.

4.4.5 Requirements of Modeling

The mDSM methodology utilizing evolutionary stability provides the software tester or designer a mechanism for decomposing modeling requirements to software components. This componentization provides an opportunity for the tester or designer to distill modeling requirements to an easily consumable matrix. The model itself provides direction for testing and design requirements. Model-based approaches provide a systematic way to identify requirements issues [45].

4.4.6 Model Validation

Model-based approaches are able to detect design and testing defects [45]. Therefore, allowing testers and designers the ability to validate the model. Given the use of the mDSM with the case ERP, we can quickly determine impacts to model. The mDSM provides a simple view of a complex software system. A way to quickly validate that the software system will meet desired criteria and where design and testing should focus.

5 Advantages of mDSM for Design and Testing Rationalization

Testing software and designing software without any iteration are challenges yet to be won! Bugs and issues are reported from software that is in use all the time. Software evolves with the aim to fix bugs, identify issues, and to add additional features. mDSM approach offers a meaningful way to optimize the number of test cases while fixing bugs and issues. In fact, one can even quantify the required number of test cases in most cases.

As stated earlier, the first step in mDSM is to decompose software components. The evolutionary metrics will highlight dependencies or interaction among components. Stated in another way, evolutionary metrics will outline the coupling and cohesion between components. The same can be extended to component-based DSM as well. If components are loosely coupled and autonomous, then testing should focus on "black box" approach, as the functionality of the component in terms of input and output remain unchanged. Once the approach is confined to black box testing, then test cases relevant to the component that has evolved or changed alone need to be executed. As mentioned earlier, one can apply appropriate techniques to quantify and identify the relationship between software evolution and test cases that need to be executed.

Just in case if the software evolution results in additional features or enhancements to functionality, then only test cases to test the additional or "*delta*" functionality need to be executed. If this results in any change to the interaction matrix, then test cases as appropriate need to be executed as well.

From a management point of view, version stability defined by mDSM approach provides a rule of thumb to decide on the effort estimation for testing. Version stability score and test efforts are inversely proportional with mDSM. The higher the version stability score, the lower the test efforts needed. Since our version stability metric identifies the percentage of source code that changed between versions, the less changes in source code must result in lower test efforts. This provides an excellent insight since it is not uncommon to find case studies where all test cases of a module are run if any change is made to the module in question and very little attention, if any, is addressed to the amount of change and its impact.

5.1 Evolutionary Stability Across Generations

The mDSM utilizing evolutionary stability provides a means to provide a holistic view across generations of software systems. This allows for the examination of a software systems adaptability over time and versions providing areas of focus for software systems designers and testers. The insight into adaptability is that the more evolutionarily stable a software system component is the less likely the system will be able to quickly adapt to change. Testers, designers, and developers need to take this into account when rapid software changes are required.

5.2 Modularization of Model

The mDSM provides the methodology to quickly decompose and modularize a software system. Individual component interaction can be mathematically realized to understand the stability or instability of a software components interaction without losing the view of the entire system. This view influences scenario-based simulation and testing, allowing for greater impact examination. Impact examination directs additional testing and design efforts.

5.3 Rationalization of Design and Test Scenarios

Rationalization of design and test scenarios can be realized through the mDSM model without making changes to the software system. Change impacts can be viewed through the model. We can easily validate and rationalize design and test scenarios based upon interactions with low stability. This provides direction for where test and design utilization should be focused.

5.4 Baseline or Benchmarking for Design and Testing

The mDSM model can be utilized as a baseline or benchmark to establish future design feasibility and benchmark testing. In relation to evolutionary stability, we can use the benchmark that provides the greatest stability, exhibiting less bugs, and having a high quality. New versions could be viewed as a modification to the benchmark. As such, we could reduce potential of regression faults and perform benchmark and comparative testing. Benchmarking and comparative testing with the mDSM improve the effectiveness of testing.

5.5 *Innovation Through Rationalization*

If we can quickly rationalize a software system and address interactions that exhibit low evolutionary stability the greater our ability to advance software to market. In order to be innovative, transformational, and disruptive, we must be able to quickly rationalize architectural and informational level decisions. By zeroing in on problem areas, we can pinpoint where testing and design efforts should be focused.

6 Conclusions

In this chapter, we studied software evolutionary stability based on Kolmogorov complexity and NCD and the mDSM based upon the extension of the DSM methodology. We introduced and defined several metrics to measure the evolutionary stability of evolving software artifacts and how they could be applied to a closed-source software system with the mDSM. We performed case studies on two open-source products, Apache HTTP and Apache Ant, and one closed-source ERP which showed that information-level evolutionary stability metrics and the mDSM can provide additional tools for software evolutionary stability measurement and the rationalization of software systems testing and design decisions

Software evolutionary stability is an important measurement for testing and design rationalization. No single metric is expected to be universally applicable to all software artifacts. The combination of architecture-level visualization of systems with the mDSM and information-level metrics assists in monitoring the software evolution process, identifying stable or unstable software artifacts, rationalization of testing and design focus, systems architectural rationalization and provides a means to be more competitive, disruptive, and innovative in the market place.

mDSM helps software development teams to identify appropriate test methodology viz. black box versus white box, as well as, the number and type of test cases to be executed as software evolves through its metrics such as version stability and NCD. By overlaying the mDSM to tune traditional testing mechanisms over multiple versions of software releases provides the opportunity for development projects to adopt a repeatable metrics-based approach to software quality. Table 9 provides a comparison of evolutionary stability metrics, and Table 10 provides a summary of analysis techniques for testing and design that can be made with the mDSM.

6.1 *Future Work*

Future work would include utilizing the mDSM with evolutionary stability for establishing baselines for design, testing, development, and deployment.

Table 9 Comparisons of evolutionary stability metrics

Authors	Reference	Type	Strength	Weakness
Nakamura and Basili	[26]	Architecture level	Capable of capturing the big picture: architecture evolution	Not suitable for measuring single module evolution
Kelly	[27]	Program level	Unified metrics suite; easy to understand by human being	Only applicable to source code and might be biased due to the fact that single measure is used in a metric
Yu and Ramaswamy	[28]	Program level	Combination of measures are used together in a metric to reduce bias introduced by a single measure	Only applicable to measuring the evolution of source code, not applicable to other artifacts, such as specification and design
Yu and Threm	This chapter	Information level	Applicable to measuring the evolution of any software artifacts, not limited to source code	Not easy to comprehend by human being, due to the fact that the measurement is performed at the binary level

Table 10 mDSM summary

Applied area	Type	mDSM
Architecture description and analysis	Dependencies	X
	Inter-component dependencies	X
	Integration points	X
	Interfaces	X
	Accepts multiple variables (metrics)	X
	Accepts parameterized variables	X
	Interaction between metrics	X
	Description of component-based system	X
Application level analysis	Problem determination	X
Change impact analysis	Change prediction	X
	Evolutionary stability	X
	Requirement change impact	X
Modularization	Undesired dependency removal	X
Quality	Prediction of defects and failures	X
Traceability and feature analysis	Link requirements analysis and design	X
	Feature location and analysis	X

Exploration of the mDSM with alternative metrics and different levels of abstraction. Additional work around the combination and complement of information-level metrics to architectural-level and program-level metrics will be also be explored. Determining impact dependent on audience: user, developer, designer, or tester.

References

1. Verganti, R, Design driven innovation, 3 Aug 2009, http://www.designdriveninnovation.com. Retrieved 20 May 2015
2. P.M. Senge, *The fifth discipline: the art and practice of the learning organization* (Doubleday/Currency, New York, 1990)
3. D. Carrington, Teaching software design and testing, in *Frontiers in Education Conference, 1998. FIE '98. 28th Annual*, vol 2, 4–7 Nov 1998, pp. 547, 550. doi:10.1109/FIE.1998.738732
4. S.S. Yau, J.J.-P. Tsai, A survey of software design techniques. IEEE Trans. Softw. Eng. **SE-12** (6), 713,721. doi:10.1109/TSE.1986.6312969
5. Introduction to OMG's Unified Modeling Language® (UML®) (n.d.). Retrieved 12 Aug 2015
6. T. Gorschek, E. Tempero, L. Angelis. On the use of software design models in software development practice: an empirical investigation. J. Syst. Softw. **95**, 176–193 (2014). doi:10.1016/j.jss.2014.03.082
7. L. Yu, D. Threm, S. Ramaswamy, Toward evolving self-organizing software systems: a complex system point of view, in *Proceedings of the 24th International Conference on Industrial Engineering and Other Applications of Applied Intelligent Systems Conference on Modern Approaches in Applied Intelligence—Volume Part II (IEA/AIE'11)*, ed. by K.G. Mehrotra, C.K. Mohan, J.C. Oh, P.K. Varshney, M. Ali, vol. Part II (Springer, Berlin, Heidelberg, 2011), pp. 336–346
8. L. Luo, Software testing techniques: technology maturation and research strategy. Class report for (2001)
9. S.G. Eick, T.L. Graves, A.F. Karr, J.S. Marron, A. Mockus, Does code decay? Assessing the evidence from change management data. IEEE Trans. Softw. Eng. **27**(1), 1–12 (2001)
10. P. Mohagheghi, R. Conradi, O.M. Killi, H. Schwarz, An empirical study of software reuse vs. defect-density and stability, in *Proceedings of the 26th International Conference on Software Engineering* (ACM Press, New York, 2004), pp. 282–292
11. T. Menzies, S. Williams, B. Boehm, J. Hihn, How to avoid drastic software process change (using stochastic stability), in *Proceedings of the 31st International Conference on Software Engineering* (ACM Press, New York, 2009), pp. 540–550
12. G. Leavens, M. Sitaraman, *Foundations of Component-Based Systems* (Cambridge University Press, Cambridge, 2000)
13. F. Dantas, Reuse vs. maintainability: revealing the impact of composition code properties, in *Proceeding of the 33rd International Conference on Software Engineering* (ACM Press, New York, 2011), pp. 1082–1085
14. E. Figueiredo, N. Cacho, C. Sant'Anna, M. Monteiro, U. Kulesza, A. Garcia, S. Soares, F. Ferrari, S. Khan, F. Filho, F. Dantas, Evolving software product lines with aspects: an empirical study on design stability, in *Proceedings of the 30th International Conference on Software Engineering* (ACM Press, New York, 2008), pp. 261–270
15. M.E. Fayad, A. Altman, An introduction to software stability. Commun. ACM **44**(9), 95–98 (2001)
16. M.E. Fayad, Accomplishing software stability. Commun. ACM **45**(1), 111–115 (2001)

17. M.E. Fayad, S.K. Singh, Software stability model: software product line engineering overhauled, in *Proceedings of the 2010 Workshop on Knowledge-Oriented Product Line Engineering* (ACM Press, New York, 2010), Article 4
18. P.E. Xavier, E.R. Naganathan, Productivity improvement in software projects using 2-dimensional probabilistic software stability model (PSSM). ACM SIGSOFT Softw. Eng. Notes **34**(5), 1–3 (2009)
19. E.R. Naganathan, P.E. Xavier, Architecting autonomic computing systems through probabilistic software stability model (PSSM), in *Proceedings of International Conference on Interaction Sciences* (IEEE Computer Society Press, Washington, DC, 2009), pp. 643–648
20. D. Grosser, H.A. Sahraoui, P. Valtchev, Predicting software stability using case-based reasoning, in *Proceedings of the 17th IEEE International Conference on Automated Software Engineering* (ACM Press, New York, 2002), pp. 295–298
21. J. Bevan, E.J. Whitehead, Identification of software instabilities, in *Proceedings of the 10th Working Conference on Reverse Engineering* (IEEE Computer Society Press, Washington, DC, 2003), pp. 134–145
22. Z. Wang, D. Zhan, X. Xu, STCIM: a dynamic granularity oriented and stability based component identification method. ACM SIGSOFT Softw. Eng. Notes **31**(3), 1–14 (2006)
23. H.S. Hamza, Separation of concerns for evolving systems: a stability-driven approach, in *Proceedings of 2005 Workshop on Modeling and Analysis of Concerns in Software* (ACM Press, New York, NY, 2005), pp. 1–5
24. S.S. Yau, J.S. Collofello, Some stability measures for software maintenance. IEEE Trans. Softw. Eng. **6**(6), 545–552 (1980)
25. S.S. Yau, J.S. Collofello, Design stability measures for software maintenance. IEEE Trans. Softw. Eng. **11**(9), 849–856 (1985)
26. T. Nakamura, V.R. Basili, Metrics of software architecture changes based on structural distance, in *Proceedings of IEEE International Software Metrics Symposium* (IEEE Computer Society Press, Washington, DC, 2005), pp. 54–63
27. D. Kelly, A Study of design characteristics in evolving software using stability as a criterion. IEEE Trans. Softw. Eng. **32**(5), 315–329 (2006)
28. L. Yu, S. Ramaswamy, Measuring the evolutionary stability of software systems: case studies of Linux and FreeBSD. IET Softw **3**(1), 26–36 (2009)
29. L. Fortnow, Kolmogorov complexity, in *Aspects of Complexity, Minicourses in Algorithmics, Complexity, and Computational Algebra* (Walter De Gruyter Incorporation, 2001)
30. C. Bennett, P. Gacs, M. Li, P. Vitányi, W. Zurek, Information distance. IEEE Trans. Inf. Theory **44**(7), 1407–1423 (1998)
31. M. Li, X. Chen, X. Li, B. Ma, P. Vitányi, The similarity metric. IEEE Trans. Inf. Theory **50** (12), 3250–3264 (2004)
32. R. Cilibrasi, P. Vitányi, Clustering by compression. IEEE Trans. Inf. Theory **51**(4), 1523–1545 (2005)
33. N. Tran, The normalized compression distance and image distinguishability, in *Human Vision and Electronic Imaging*, vol. XII (2007), 64921D
34. T. Arbuckle, A. Balaban, D.K. Peters, M. Lawford, Software documents: comparison and measurement, in *Proceedings of the 19th International Conference on Software Engineering & Knowledge Engineering* (Knowledge Systems Institute Graduate School, Skokie, 2007), pp. 740–745
35. T. Arbuckle, Visually summarizing software change, in *Proceedings of the 12th International Conference on Information Visualisation* (IEEE Computer Society Press, Washington, DC, 2008), pp. 559–568
36. T. Arbuckle, Measure software and its evolution using information content, in *Proceedings of the Joint International and Annual ERCIM Workshops on Principles of Software Evolution and Software Evolution Workshops* (ACM Press, New York, 2009), pp. 129–134
37. T. Arbuckle, Studying software evolution using artefacts' shared information content. Sci. Comput. Program. **76**(12), 1078–1097 (2011)

38. T. Arbuckle, Measuring multi-language software evolution: a case study, in *Proceedings of the 12th International Workshop on Principles of Software Evolution and the 7th Annual ERCIM Workshop on Software Evolution* (ACM Press, New York, 2011), pp. 91–95
39. S.D. Eppinger, T.R. Browning, *Design Structure Matrix Methods and Applications*, 1st edn, vol. 1 (MIT Press, Cambridge, 2012), number 0262017520
40. T.R. Browning, Applying the design structure matrix to system decomposition and integration problems: a review and new directions. IEEE Trans. Eng. Manage. **48**(3), 292–306
41. A. Thurimella, S. Ramaswamy, On adopting multi-criteria decision-making approaches for variability management in software product lines, in *First International Workshop on Requirements Engineering Practices On Software Product Line Engineering, 16th International Software Product Line Conference*, Salvador, Brazil, 2–7 Sept 2012
42. A. Engel, T.R. Browning, Designing systems for adaptability by means of architecture options. Syst. Eng. **11**, 125–146 (2008). doi:10.1002/sys.20090
43. T.B. Callo Arias, P. Spek, P. Avgeriou, A practice-driven systematic review of dependency analysis solutions. Empirical Softw. Eng. **16**(5), 544–586
44. R.R. Yager, Intelligent control of the hierarchical agglomerative clustering process. IEEE Trans. Syst. Man Cybern. B **30**, 835–845 (2000)
45. D. Aceituna, Do. Hyunsook, G.S. Walia, S.-W. Lee, Evaluating the use of model-based requirements verification method: a feasibility study, in *Empirical Requirements Engineering (EmpiRE), 2011 First International Workshop on*, 30 Aug 2011, pp. 13–20
46. Frequently Asked Questions. *Apache Ant*. N.p. (n.d.). Web. 6 Dec 2015
47. Apache Release History, *Apache Release History*. N.p. (n.d.). Web. 6 Dec 2015
48. A. Causevic, D. Sundmark, S. Punnekkat, An industrial survey on contemporary aspects of software testing, in *Software Testing, Verification and Validation (ICST), 2010 Third International Conference on*, 6–10 April 2010, pp. 393, 401. doi:10.1109/ICST.2010.52
49. M. Venkataraman, Testing services, 29 June 2015. Retrieved 2 Aug 2015

Testing as a Service

Pankhuri Mishra and Neeraj Tripathi

Abstract Testing as a Service (TaaS) [1] is gaining acceptance in software engineering industry for it finds outsourcing of testing a viable option for both, adhering to software production timeline and making testing cost-effective while not making compromise in product quality. In consequence, few companies have started offering tools automating TaaS process. This chapter highlights on this upcoming concept and presents a generic model explaining the steps involved in TaaS. It also explains use of cloud technology for TaaS implementation. This aims at effective utilization of resources while delivering a testing service. In addition, a pricing model for test outsourcing is proposed.

Keywords Taas · Test optimization · Testing as a service · Taas on cloud · Outsourcing · Test outsourcing · Test service · Taas enablers · Benefits of Taas · Taas accelerators · Traditional testing to Taas · Taas limitations · Taas disadvantages · Taas architecture · Taas working model · Taas framework · Taas demo · Taas example · Taas infrastructure · Industry case study in Taas · Taas cost model · Pricing test service · Industry leaders in Taas · Taas provider · Taas consumer

1 Introduction

Software [2] is a set of machine-readable instructions to perform specific operations. With various devices so embedded in our lives, we are more dependent on software than ever before. From the alarm clock that wakes us up, transport that

P. Mishra · N. Tripathi (✉)
Microsoft India, Hyderabad, India
e-mail: neeraj.tripathi@microsoft.com

P. Mishra
e-mail: panmis@microsoft.com

H. Mohanty et al. (eds.), *Trends in Software Testing*,
DOI 10.1007/978-981-10-1415-4_7

helps us commute, ATMs for money withdrawal, smart phones for communication to television for entertainment, almost everything runs on the software.

Software plays such an important role in our life, so it is of utmost importance to ensure software quality. Testing is the process of verifying the quality of any product. Let us consider an example from the textile industry. While buying clothes, we expect them to be durable, have appropriate size and long-lasting color. Before selling a cloth, it is the seller's responsibility to test these basic features of the cloth or in short, the seller is responsible for the quality of the cloth.

Software testing is the process where software engineers must assure the quality and correctness of any software before selling it to the customers. Any compromise in software testing can cause huge monetary loss, sometimes even loss of human lives; e.g., In Scotland, a Chinook helicopter crashed due to a software defect and all 29 passengers got killed [3]. Arianne-5 rocket was set to deliver a payload of satellites into Earth's orbit, but a software defect caused 27 seconds delay and eventually resulted in the mission failure. More than 370 million dollars were lost in this mission [4]. A study by the National Institute of Standards and Technology, USA, found that every year software defects cost the US economy $59.5 billion [5]. To elaborate software testing, let us consider a flight-booking portal. It is the portal owner's responsibility to ensure correctness of the information on the portal.

Each software development company hires, especially skilled testers for testing products. Testing is about 25 % of software life cycle [6]. It is the tester's responsibility to

1. Set up the test environment by positioning the right set of machines. For example, for the flight-booking portal, we need various devices like laptop, tablet, and mobile of various screen sizes to make sure user interfaces (UI) work for all the devices.
2. Write expected behavior for various features and tests to verify. This phase is termed as creating a test plan. For the flight-booking portal, we ensure that all the flight information including seat availability is correct. Based on given information, a test plan should cover test cases to verify all the functionalities the portal provides.
3. Once a product is ready, it has to be tested under various setups and environments. For the flight-booking portal, we need to test its usability, scalability, and performance on various device setups.

Software testing can be of various types [7], a tester mainly considers following testing types while creating test plans:

1. **Functional Testing**—Test the software output and verify whether it is the same as mentioned in the requirements specification.
2. **Integration testing**—Test various software modules to verify combined functionality after integration. For the flight-booking portal, we have various modules like portal UI to display the flight information and database to store flight information. It is important to verify that the information displayed on the portal is same as that stored in the database.

3. **Sanity testing**—In sanity testing, we test just the basic feature and ignore the secondary features. Sanity testing is the subset of functional testing and we perform sanity testing, when we do not have enough time for complete testing. For flight-booking portal, displaying correct flight name and seat availability are primary features. In sanity testing, we might overlook if some flights are missing from the portal but there should be no wrong information.
4. **Regression testing**—Test the complete product after modification of a module. For a flight-booking portal, it is important to ensure that addition of information like adding a new flight detail or modification to a module should not risk the correctness of the portal.
5. **Load/Stress testing**—Test under heavy loads to determine at what point the system's response time degrades or fails. This is accomplished by making a system respond to both fast evolving data stream and large volume of data as inputs. Incase of a flight-booking portal, at any point of time, if thousands of users are trying to book flights, the portal should be able to handle them with reasonable response time.
6. **Usability testing**—To test whether a new user can use the software easily. The product should have help or tips whenever needed. A flight-booking portal needs to have help and warning notes wherever needed.

As discussed above, in traditional testing model, testers set up a testing environment, create test plans, and execute the test plans. We discussed various testing types. Overall, the model looks promising. However, the traditional testing model has a few challenges such as:

1. **High cost**: Usually, the cost associated with building test expertise, buying or creating best automation tools, and maintaining test infra is very high especially for small and medium businesses. Moreover, we do not use test resources during complete software product life cycle. At times, we do not perform any testing but need to invest in the test infrastructure maintenance cost. For a flight-booking portal, we do not need the testing devices all the time, but the maintenance associated with them is a recurring cost.
2. **Lack of Resource flexibility**: Traditional models do not allow to staff up testing experts quickly when needed and release when not required. There are certain testing experts that software companies need during stress testing, but may not need for other testing types. As a result, you will have to hire a person permanently for limited work. In case of a flight-booking portal, you might want to test it against security threats and hacks and need to hire specialized testers to perform security testing. These testers are helpful only after the product is ready.
3. **Expense for Advanced Test Environments**: Advanced tools, latest devices, and platforms need frequent updates and are expensive. In case of a flight-booking portal, with new devices and versions releasing almost every month, it becomes a bottleneck to have all the latest devices in-house for testing purpose.

4. **Lack of specialized test expertise**: There is a lack of specialized testers. Testers may need a good amount of ramp-up time for different testing tools and technologies. Similarly, testers actively working on security domain have the knowledge of the latest security threats and possible future attacks. The challenge is to hire these testers for small yet very significant part of testing.
5. **Absence of fully Automated Testing**: With a growing business, it becomes difficult to supply a fully managed and automated testing. Usually, managing manual testing for the large applications becomes more error prone. For a flight-booking portal, manual UI verification on all the possible devices, including laptops, tablets, and mobiles is very difficult.

Looking at the best practices and challenges of the traditional testing model mentioned above, is there a way to continue getting the benefits of the traditional testing model and transfer the challenges to someone else experienced in that field?

For example, a cloth manufacturer can focus on manufacturing and hire an external agency to verify the cloth quality. In this way, manufacturers can create a variety of clothes and the external agency that primarily works on ensuring the cloth quality can provide the quality assurance.

With respect to the flight-booking example, a company can let developers focus on building better UI and efficient databases and hire an external agency to verify the quality of the portal. Also, such agencies can share the devices and testers available among them and get a better return on investment (ROI).

Testing as a service (TaaS) is a model for outsourcing testing responsibilities to external and specialized providers. Typically, product owners or the service consumers do some testing in-house that require code knowledge and they require external expertise in other related areas. Testing requirements become complex with growing applications. The cost of automation tools and the skilled testers can be expensive. TaaS becomes a good business case because it gives a good ROI for optimal usage of skilled resources and tools for testing.

In this chapter, we cover basic concepts on TaaS and its advantages and limitations in detail. Section 2 presents TaaS architecture and Sect. 3 proposes a framework that enables TaaS. Section 4 gives an imprint of its behavior through an example. Section 5 proposes TaaS pricing model. The section gives details on how to measure the efforts and cost for TaaS. Section 6 deals with TaaS on cloud. It talks about key participants and advantages of TaaS on cloud. The section presents a strategy for making TaaS working on cloud. The chapter ends in the next section with a concluding remark.

2 Testing as a Service

"As a Service" solutions are becoming popular as these enable better ROI without infrastructure investments. In testing, advanced testing efforts require expertise, large amount of resources, and expensive testing tools. Pay-per-use models allow software companies to get what they need and when they need it. As a result, TaaS solutions are becoming increasingly popular. Let us understand what makes TaaS delivering a testing service. In this context, next we present an infrastructure that enables to deliver TaaS.

2.1 TaaS Architecture

TaaS is a service provider and consumer interaction-based model. Service consumer places a request to provider and provider fulfills the request. Priorities are typically easy user interface, data security, and service availability and most importantly cost. Figure 1 shows the TaaS architecture.

Let us look at each of these components in detail to understand the components and the interaction better.

2.1.1 Service Provider

A service provider provides testing services to any company. The key services include test consultation, planning, automation, execution, and management. Figure 2 illustrates the responsibilities of a service provider.

Some of the key driving objectives while planning the testing strategy are:

1. Quality: A number of users, at times millions, interact with the applications or products. With such high usage and dependency on the data correctness, quality

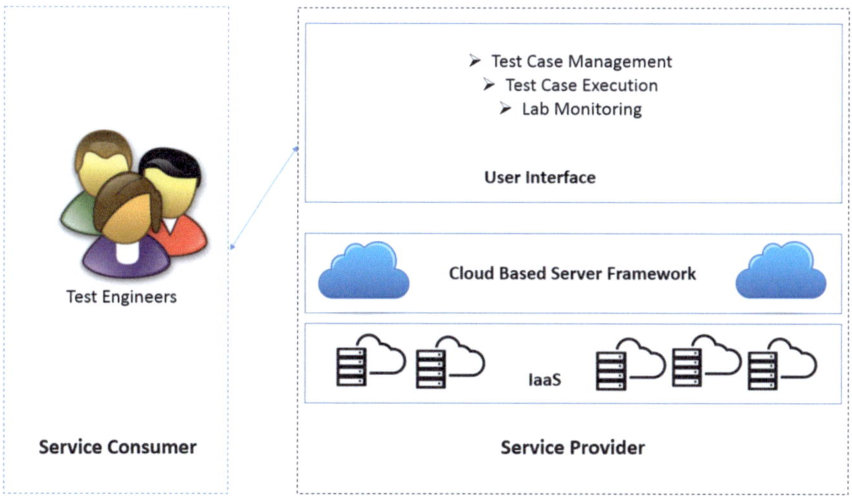

Fig. 1 TaaS architecture

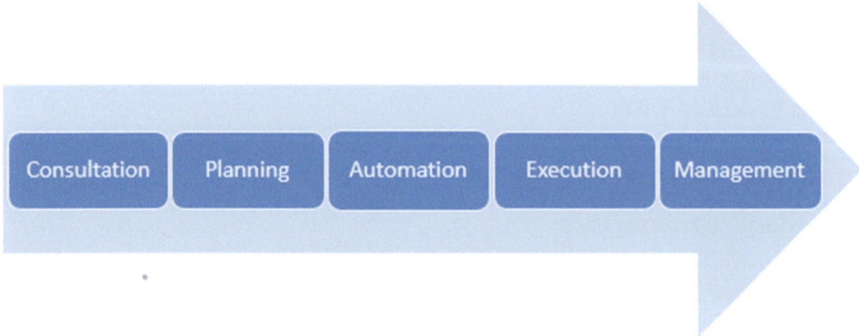

Fig. 2 Responsibilities of a service provider

is the top priority. Service providers create real-world environments to test applications. Quality is the key objective for any TaaS provider.
2. Efficiency: Service providers focus on reducing cost by centralized testing, tools optimization, and overall resource effectiveness. This results in faster testing and early identification of issues.
3. Revenue: Service providers focus on building expertise in-house. This helps to generate more revenues and expand the testing domains.

Figure 3 shows the responsibilities of a service provider.

Time invested in planning helps in a long way. It avoids any confusion later between the service provider and the consumer. The service provider needs to take care of following factors while planning.

1. Identify all the platforms on which the product is applicable.
2. Document the business and functional requirements.
3. Evaluate testing efforts.
4. Plan for revising testing efforts on regular interval.
5. Plan a postmortem meeting after every release.

Fig. 3 Responsibilities of a service provider

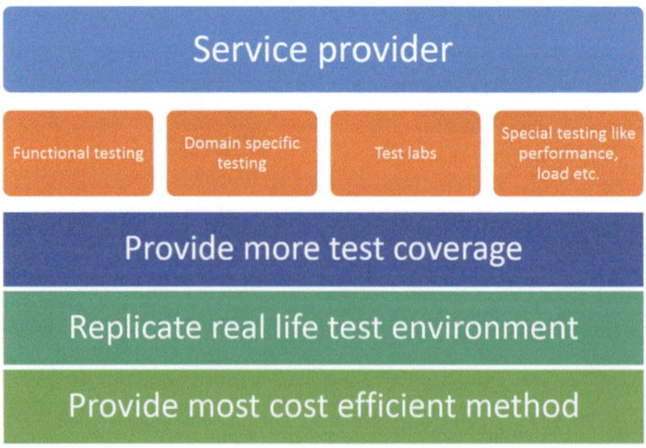

6. Identify the effective communication mechanism. For example, communicate bugs or issues through e-mails.

2.1.2 Service Consumer

A service consumer uses the outsourcing model to get a service and pay for it accordingly [8]. Automated regression testing, security testing, performance testing, monitoring and testing of cloud-based applications, and testing of major ERP software such as SAP are most suited candidates for TaaS model since they require expertise and infrastructure. TaaS is also a promising solution for smaller business houses, who do not have the resources to invest in testing skills and want to focus entirely on product development.

When a service consumer moves to TaaS model, the consumer needs to establish a well-defined process, identify the specific testing area where TaaS is applicable, and align development teams to TaaS model. Therefore, preparing a high-level transformation road map is important in achieving the benefits of TaaS model.

A consumer cannot simply jump to an end-point but must plan series of phases. These include

1. Identify current test model and verify whether to move complete testing or any specific tests to TaaS.
2. Select the service provider based on the testing requirements for better ROI.
3. Start the transition with smaller release cycles and pilot programs.
4. Clearly define the testing requirements before release cycles.

Figure 4 highlights role of consumers in TaaS model.

After the pilot program, the consumer assesses the effectiveness of the current model. The service consumer considers following points before deciding to continue with the service provider:

- Communication is the key so it is important to establish an engagement model.
- Identifying how well the service provider aligns to the process.
- A mechanism to monitor the performance of services through service level agreements (SLAs).
- The payment mechanism suits the consumer needs with no hidden cost.

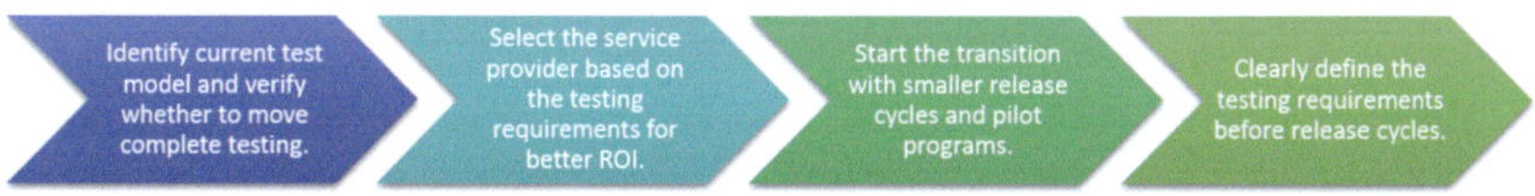

Fig. 4 Role of a consumer in TaaS

2.1.3 Interaction Between a Service Provider and a Consumer

Service consumer initially shares the test requirements with the provider. The requirements include product details, type of testing needed, and timelines. Based on the requirements, service provider calculates the total cost and creates a test plan. Based on prices, the consumer can negotiate with the provider regarding the total cost. The consumer can also choose among the services offered by the provider. After finalizing the contract, the provider performs the testing and shares the results with the consumer. Figure 5 shows the flow of interaction between the service provider and the consumer.

Service provider interacts with the consumer typically over e-mails. Initially, service provider should have a few in person meetings and communication with the consumer to speed up the process. It also helps to establish an initial trust between the provider and the consumer. However, the most common model is service providers having their Web portals. Service consumers can use these portals for any kind of communication.

2.2 *TaaS Enablers*

Now, we understand the TaaS architecture and various roles involved in TaaS. At this time, let us look at the factors favorable to TaaS:

1. In the past, software companies looked at testing as investments done for product quality. Today, software companies explore options to reduce testing costs and consider external experts to perform testing.

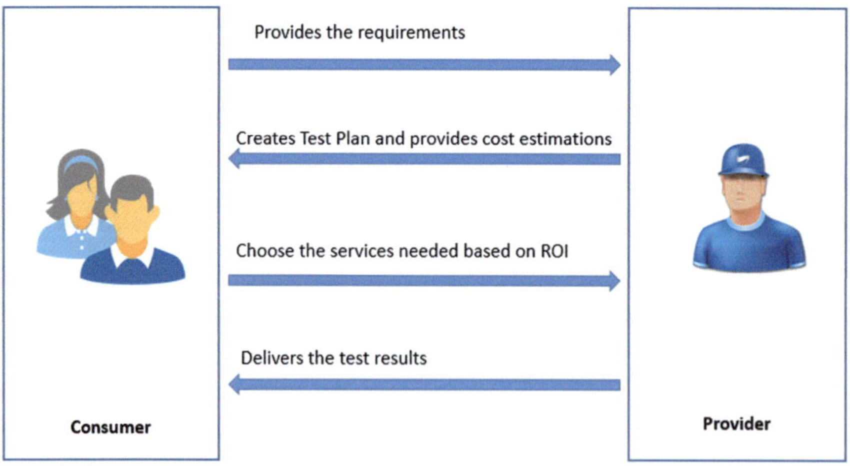

Fig. 5 Interaction between consumer and provider

Table 1 Comparison of traditional model and TaaS

	Traditional	TaaS
Supported methodologies like agile/scrum/waterfall	Available	Available
Test environments	Manual	On demand
Test data	Manually generated	Dynamically sanitized
Test framework	Manual + automated	Automated
Test tools	Manually purchased	On demand
Test documentation	Not required	Compulsory

2. TaaS framework has ability to provide testing services to several clients in parallel on optimizing usage of testing resources.
3. Earlier, due to network bandwidth constraints, software companies preferred to have co-located product development and testing teams. Improvements in network quality over the past few years enable teams to work seamlessly from different locations.

Table 1 illustrates the difference between traditional model and TaaS.

TaaS allows the service consumer to focus on its core strength rather than putting efforts in testing. The service consumer does not invest any time and efforts to create and maintain test environments and resources. Testing should be a black box to the consumer. Considering all these factors, TaaS looks promising. TaaS is certainly a big leap in test outsourcing. Table 1 presents a comparison between traditional testing model and TaaS model. From six different criterias that are vital to testing, the strength of TaaS is observable from the comparison drawn in the table.

2.3 *Moving from Traditional Testing to TaaS*

As described above, TaaS looks promising. Let us consider the efforts involved in moving from traditional testing to TaaS. This transition [9] from traditional testing to TaaS involves positive risks. Let us understand the initiating point, key steps and challenges involved in moving from traditional testing to TaaS. Figure 6 outlines the systematic flow [10].

- Identify goals: The service consumer needs to understand the goal when moving to TaaS. This goal can be to focus on core, cost savings, service of experts, etc.

Fig. 6 Transition from tradition model to TaaS

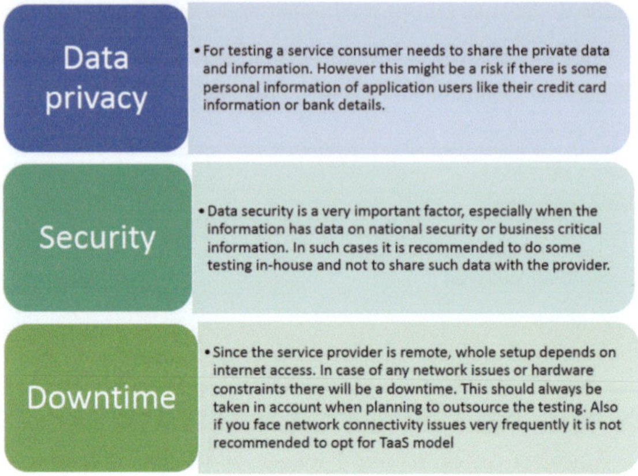

Fig. 7 TaaS limitations

- Document Requirements: A service consumer needs to do a thorough job in listing requirements. This includes putting details of the components to be tested, supported devices, technology preferences if any, performance considerations, security, scalability options, etc.
- Service provider selection: Based on the requirements, a service consumer selects a service provider who is an expert in testing domain. The service consumer checks the past record of service providers, calls for proposals from service providers, and compares the various proposals received to make a decision.
- Performance checkpoint: A service consumer needs to assess the performance of a service provider at regular intervals on defined parameters. The service consumer plans corrective actions based on the assessment findings. Then after availing the service, the consumer shares constructive feedback with the service provider.

2.4 TaaS Limitations

As described in preceding sections, TaaS is beneficial to consumers and providers. We looked at TaaS enablers in Sect. 2.4. However, TaaS also has a few limitations.

Figure 7 gives an overview of the major limitations with TaaS.

3 TaaS Infrastructure

We have discussed TaaS concepts in detail. In this section, we are presenting a generic framework for TaaS. We are building this framework to achieve complete TaaS implementation from requirements gathering to sharing test results including

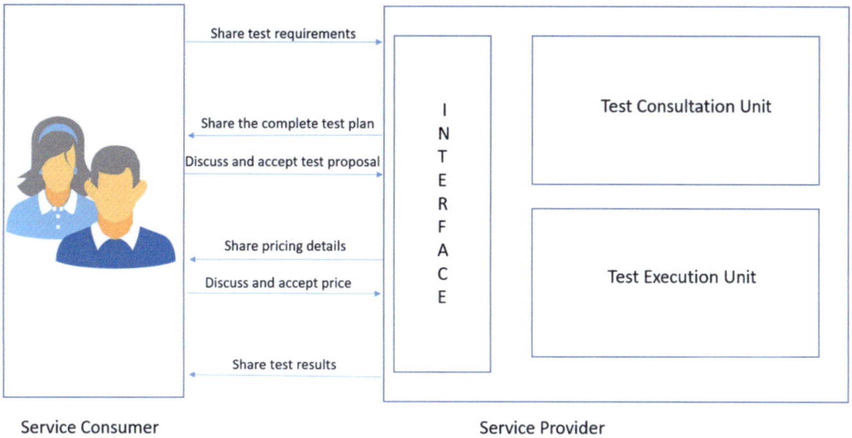

Fig. 8 TaaS framework

test plan creation, pricing, test execution, and interactions between a service provider and a consumer.

Figure 8 gives an overview of the proposed TaaS framework.

As shown in Fig. 8, the service provider exposes an interface to the service consumer. The consumer places a request, by sharing all the testing requirements through the interface. The service provider internally processes the requirements and provides a test plan. After the service consumer confirms the test plan, the service provider shares pricing details. Once the service consumer accepts the pricing, the service provider executes the test cases and shares the test results with the consumer.

The proposed TaaS framework has three main components: interface, test consultation unit, and test execution unit. Let us look at these units in detail.

3.1 Interface

The service provider exposes interface unit to a service consumer. The unit contains the user interface and keeps internal processing abstract. Therefore, the consumer need not worry about the internal processing.

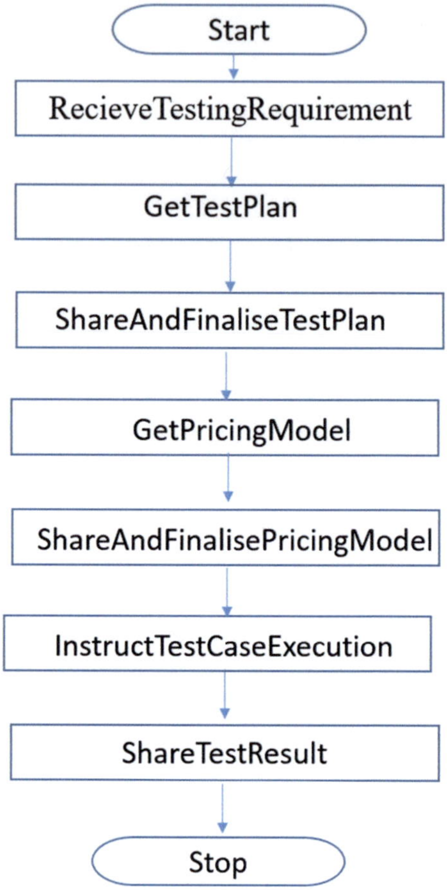

Fig. 9 Flowchart for interface unit

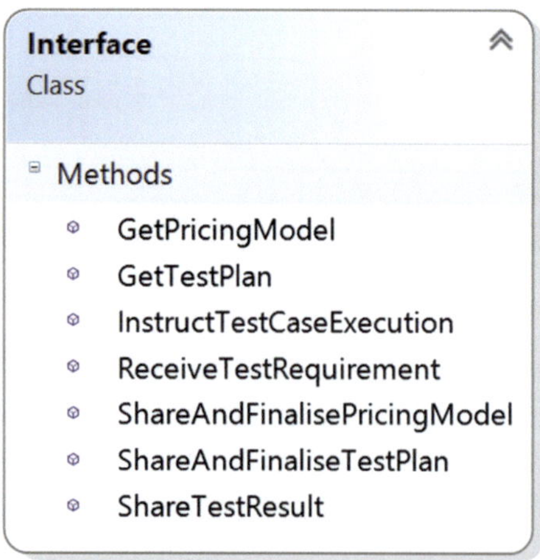

Fig. 10 Class diagram for interface

As shown in Fig. 9, the service provider receives test requirements from the service consumer through interface using ReceiveTestingRequirements method. After receiving the test requirements, interface invokes GetTestPlan method to get the test plan from test consultation unit. Through ShareAndFinaliseTestPlan method, interface shares this test plan with the service consumer. Once the consumer approves the test plan, interface invokes GetPricingModel method to get the pricing from test consultation unit. Interface shares this pricing model with the service consumer in ShareAndFinalisePricingModel method. After the consumer agrees upon the pricing model, interface invokes InstructTestCaseExecution method from test execution unit. After test execution, interface invokes ShareTestResults method to share the test results with the service consumer.

Figure 10 shows the class diagram for interface.

3.2 Test Consultation

Test consultation unit mainly creates the test plan based on the consumer's requirements and generates the price accordingly. Figure 11 shows an overview of test consultation unit.

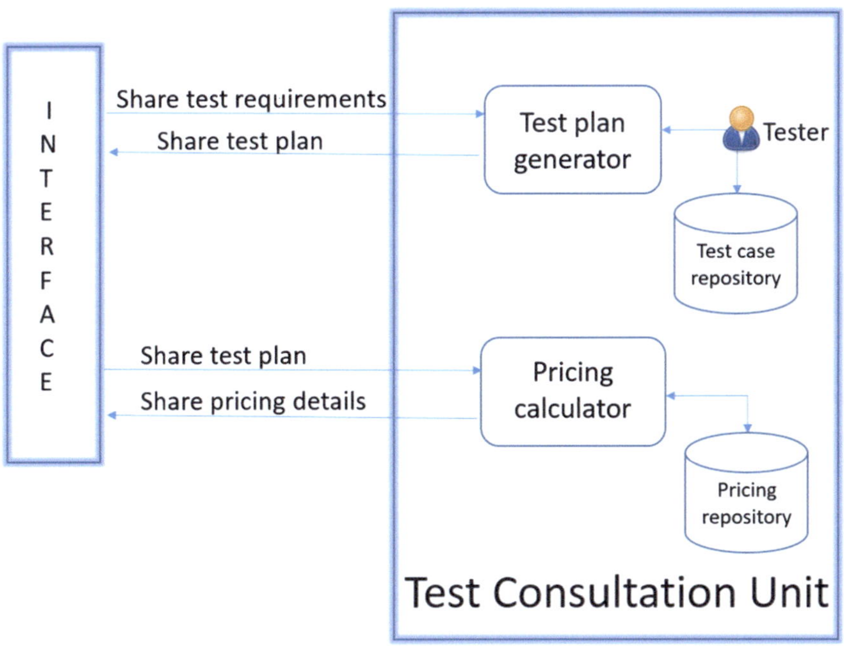

Fig. 11 Test consultation unit

As shown in Fig. 11, test consultation unit does not directly interact with the consumer. It gets the instructions and details from the interface.

Test consultation unit mainly offers two functionalities:

Test plan generator: The framework provides the support for test plan creation based on requirements specified by the consumer. It fetches test cases from an existing repository and updates the test plan. When the test plan generator does not find any test cases in Test case repository, it updates the test plan with no test cases found message.

Pricing calculator: The framework provides a generic support to calculate the price. Section 5 in this chapter provides more details on how the service provider and the consumer decide price. The agreed upon test service cost is stored in Pricing repository.

Let us look at these functionalities in detail.

3.2.1 Test Plan Generator

Figure 12 shows the flowchart for test plan generator. The interface first invokes the ProcessTestRequirement method. This method reads the test requirements shared by a consumer and converts the requirements into a standard format understandable by the framework.

Based on the requirements, the GenerateTestScenarios method generates a set of user test scenarios. GetScenarioDependents method identifies the devices (phone/tablet/laptop, etc.) and the operating system versions on which the product needs to be tested. Simultaneously, FetchApplicableTestCases method gets required test cases from test case repository. There can be manual or semiautomated

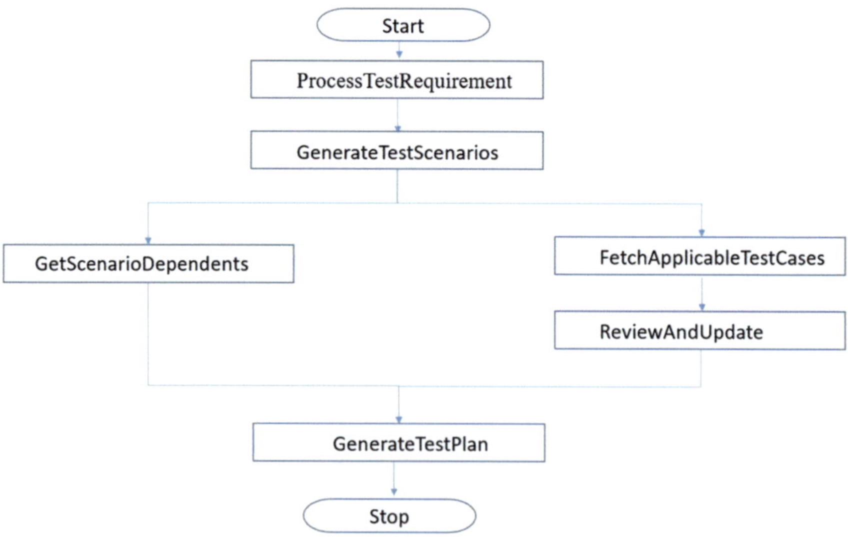

Fig. 12 Flowchart for test plan generator

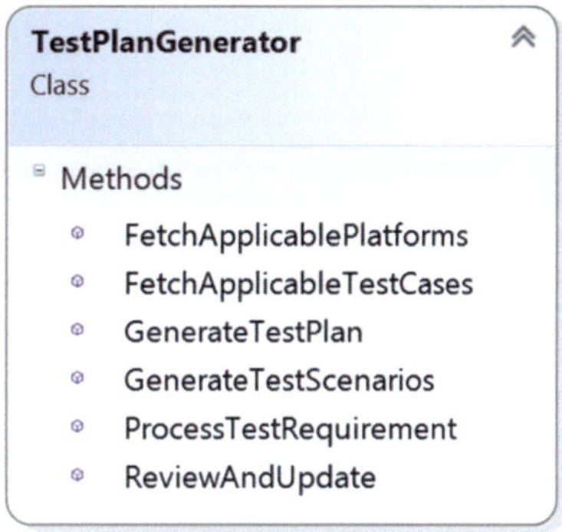

Fig. 13 TestPlanGenerator class

(ReviewAndUpdate) intervention to modify/add test cases if required. On finalization of test cases GenerateTestPlan method is invoked to create final device-dependent test plan. Figure 13 shows the class diagram for TestPlanGenerator class.

3.2.2 Pricing Calculator

Figure 14 shows the flowchart for Pricing calculator.

Pricing calculator interacts with the interface unit of Fig. 11. The framework provides a support to process a test plan and identify testing units. Section 5

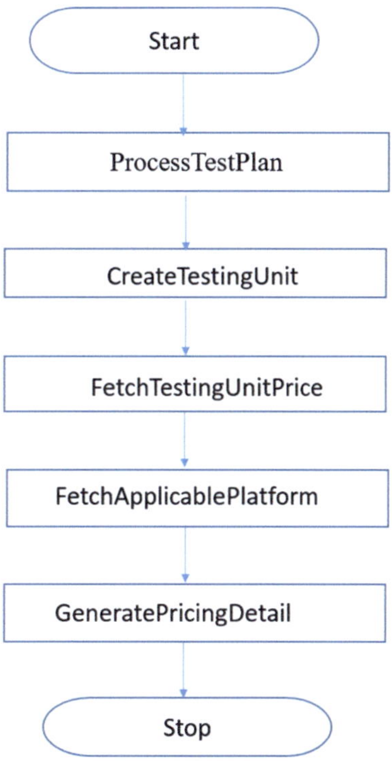

Fig. 14 Flowchart for Pricing calculator

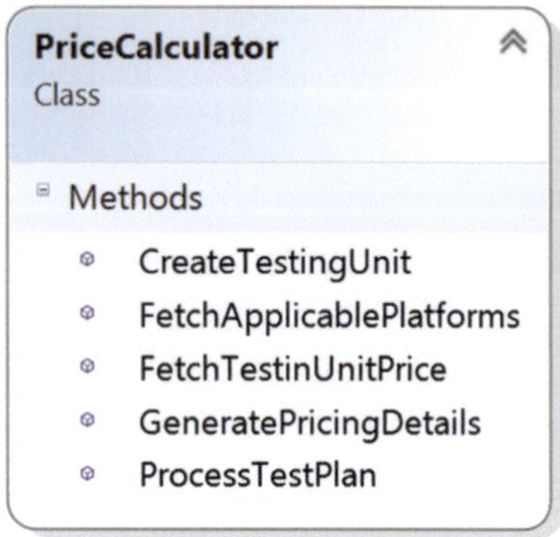

Fig. 15 PriceCalculator class

explains testing units. After identifying testing units, interface fetches the price per unit and gets its dependents. As mentioned earlier in test consultation unit, these dependents represent the devices and software platforms on which an application needs to be tested. In case a platform is not available with a service provider, the cost of building/hiring new platform is also included. After all these calculations, GeneratePricingModel method generates the final price for testing of an application.

Figure 15 shows the class for price calculation unit.

3.3 Test Execution

Test execution unit is responsible for automating or reusing the test cases, executing required test cases, and sharing test results based on the test plan shared by the interface. Figure 16 shows an overview of test execution unit.

Let us look at the operations performed by test execution unit in detail:

- *Automate test cases*: The framework supports to add new test cases and modify or reuse the existing. As described in Sect. 3.2, if a new test case it added, tester needs to automate it and add to the test case repository.
- *Execute test cases*: The executor fetches all the applicable test cases and executes them one by one. It understands various test case formats. Accordingly, the executor executes the test cases. The executor takes care of all the setup required for a test case.
- *Generate the test results*: Test result generation is the most critical part of the framework. In the report generator, we implement the logic to convert the result of test cases into a readable and user-friendly format. The report uses easy English or local language, and for better readability, inserts tables and graphs.

In this section, we present a framework that supports above three operations. This is a generic framework and we can add any number of test cases to it.

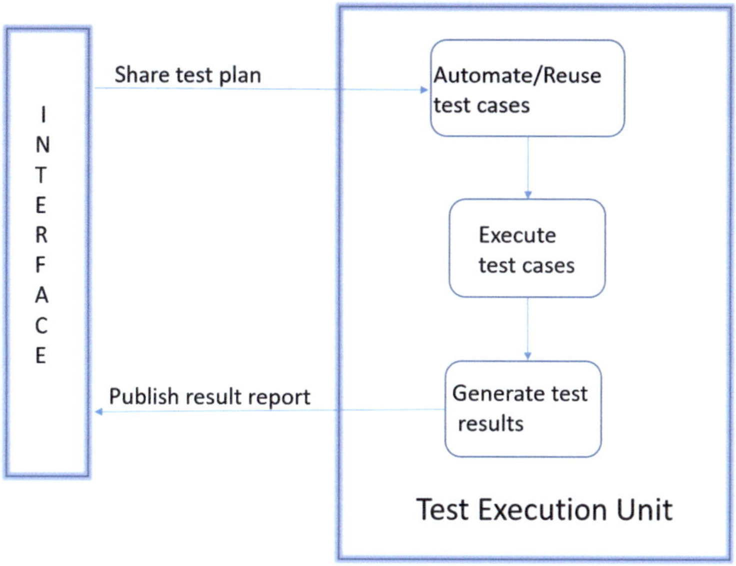

Fig. 16 Interaction of interface and test execution unit

The below TaaS framework is implemented using C#; users can implement similar solutions using other languages as well.

3.3.1 Automate Test Cases

For automating a test case, we ensure that all the test files follow a well-defined convention, as the generic framework understands the files based on this convention. New test case adds a new test class instead of modifying the existing one. Tester needs to define a class, inherited from ITestCase called as test class. ITestCase specifies the methods that a test class needs to implement. Figure 17 shows ITestCase.

Fig. 17 Class for ITestCase

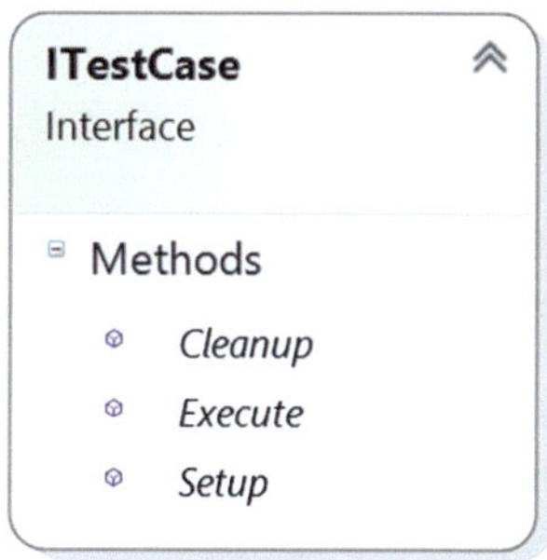

ITestCase interface has three methods named as Setup, Execute, and Cleanup. First, we execute the Setup method. The method performs setup steps such as reading input data, establishing network connection, or reading database. Execute method contains test case logic. In Execute method, we perform the testing steps. Cleanup method releases all the resources used in previous two steps. In addition, Cleanup method deletes temporary files and databases or closes all the network and database connections. A test case class needs to inherit from ITestCase.

3.3.2 Execute Test Cases

After fetching existing and the newly added test classes, we need to write methods that support the execution of these test classes. For this purpose, we implement a test case executor shown in Fig. 18.

The executor fetches all the test class and executes them on the test machine, one by one. Figure 19 shows code snippet for the method that fetches all the test classes.

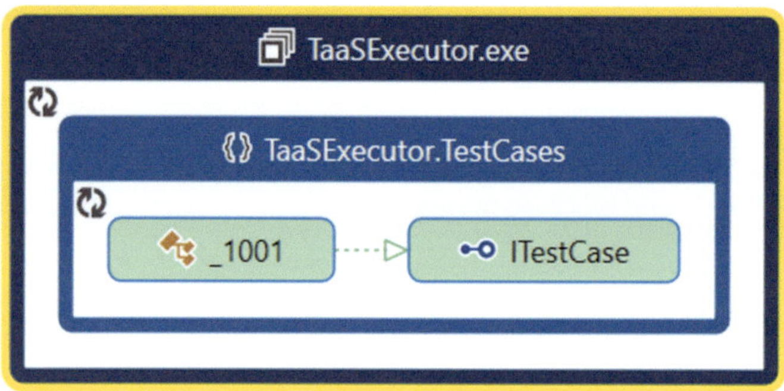

Fig. 18 Test case executor

```
    /// <summary>
    /// Gets all the test cases of given category
    /// </summary>
    /// <param name="TestCaseCategory"></param>
    /// <returns></returns>
    1 reference
    public static SortedDictionary<string,ITestCase> GetAllTestCases(TestCase TestCaseCategory)
    {
        var scenarios = new SortedDictionary<string, ITestCase>();
        var assembly = Assembly.GetExecutingAssembly();
        Type typeIScenario = typeof(ITestCase);
        foreach (Type type in assembly.GetTypes())
        {
            if (type.IsClass && type.IsPublic && typeIScenario.IsAssignableFrom(type))
            {
                if(TestCaseCategory == TestCase.All)
                scenarios.Add(type.Name, assembly.CreateInstance(type.FullName) as ITestCase);
            }
        }
        return scenarios;
    }
}
```

Fig. 19 Method to get all the test classes

```
var TestCases = GetAllTestCases(TestCase.All);
foreach (string testcase in TestCases.Keys)
{
    ITestCase currentTestCase;
    TestCases.TryGetValue(testcase, out currentTestCase);
    currentTestCase.Setup();
    currentTestCase.Execute();
    currentTestCase.CleanUp();
}
```

Fig. 20 Method to execute all the test classes

This method reads all the C# classes that inherit from ITestCase. As shown in Fig. 19, scenarios object fetches all the test classes with respect to a test case category and returns all scenarios pertaining to a given test case.

Figure 20 shows the basic functionality of test case executor. It first fetches all the test classes and then executes them one by one as shown in Fig. 20. We can extend the same method to execute test classes across multiple devices.

3.3.3 Generate Test Results

Now, we have a basic understanding of creating and executing test classes. After executing all the test cases, it is very important to generate test results. The service provider needs to share these test results with the consumer. The results include number of tests executed, outcome, error details for any failure, and time taken, etc.

Code shown in Fig. 21 generates test results.

We use GetAllTestCase method implemented in Fig. 19 to get list of all the test cases. We read results from all test cases and store in an HTML Object. We are using .Net libraries in the GenerateReport method to create HTML file. Figure 21 presents a code snippet for generation of test result document.

In this section, we have a presented an abstract implementation of the proposed generic framework that we can use for TaaS solution. We can add test cases in the test case repository and extend the executor to execute tests on various devices. Accordingly, we need to modify the test result generation. In next section, we present a case study illustrating steps involved in availing a test service by making use of the proposed framework.

```
/// <summary>
/// This function generate the report for all test cases :
/// </summary>
1 reference
static void GenerateReport()
{
    var TestCases = GetAllTestCases(TestCase.All);
    var htmlTable = HTML.CreateTable();
    var headerRow = new List<String> { "ID", "Name", "Total Tests", "Passed", "Failed",
                                        "Details", "Time Taken (in secs)", "Owner" };
    htmlTable.AddRow(headerRow);

    foreach (string testcase in TestCases.Keys)
    {
        ITestCase currentTestCase;
        TestCases.TryGetValue(testcase, out currentTestCase);
        string testID = currentTestCase.GetID();
        string testName = currentTestCase.GetName();
        string totalTests = currentTestCase.GetTotalTests();
        string passedTests = currentTestCase.GetPassTests();
        string failedTests = currentTestCase.GetFailedTests();
        string details = currentTestCase.GetDetails();
        string timeTaken = currentTestCase.GetTimeTaken();
        string ownerName = currentTestCase.GetOwnerName();
        var resultRow = new List<String> { testID, testName, totalTests, passedTests,
                                            failedTests, details, timeTaken, ownerName};
        htmlTable.AddRow(resultRow);
    }

    htmlTable.SaveHTMLReport();
}
```

Fig. 21 Method to generate test results

4 An Experiment

In Sect. 3, we have presented a generic TaaS framework. There we discussed on all the components of the framework. In this section, we discuss on working of the framework through an example.

4.1 A Case Study

Let us take flight-booking portal example to understand the application of TaaS concepts in the real world. The flight-booking portal has multiple flight carriers operating various domestic and international flights. A user comes to the portal and gets details for all the flight carriers at a single glance.

The flight-booking portal has to add new flight carriers. Let us look at the Fig. 22 to understand addition of new flight carrier to the portal.

The flight-booking portal receives the flight details from various flight carriers, performs validations, applies required transformations, converts the details to standard XML files, and moves the data to database post-validation. In the current

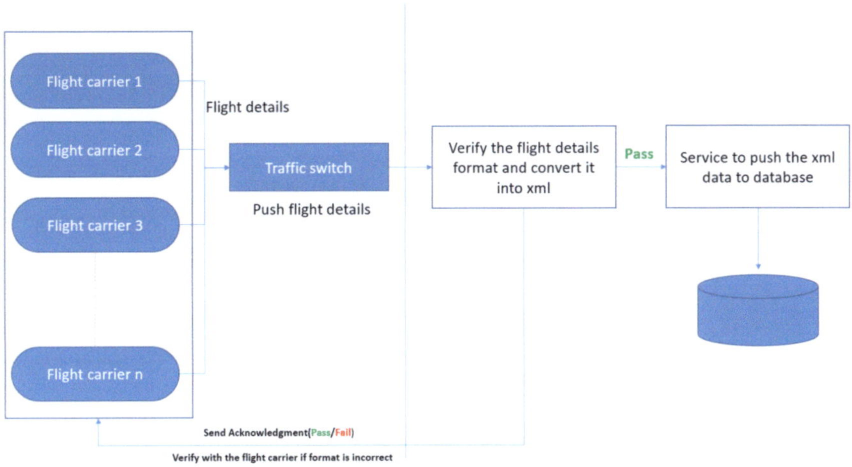

Fig. 22 Adding a new flight carrier to flight-booking portal

scenario, adding a new flight carrier requires testing coordination between the flight carrier and the company (flight-booking portal). The company needs to maintain a dedicated team to coordinate with the flight carrier. The team creates required test plan and executes the test cases as per the test plan.

Suppose, there is a service provider who takes care of testing framework to add a new flight carrier. Figure 23 shows the interaction between the portal and the service provider. Service consumer here is the flight-booking portal.

As shown in Fig. 23, service consumer (the portal) shares the testing requirements. The service provider assesses the requirements and identifies the components involved to fulfill this requirement. The pricing model is decided. We discuss TaaS pricing model details in Sect. 5. Up to this time, both service provider and the service consumer understand the requirements and agree on the cost.

From here on, the service provider coordinates with new flight carrier for all the required testing. Now the company can focus on its core to enrich its database and provide better user experience.

Figure 24 explains the interaction between flight carrier and service provider.

As shown in Fig. 24, the service provider validates the flight details sent by the carrier. The provider does validations through automated test cases. As and when, we need to add a new flight carrier; the service provider fulfills the testing needs. Now, the flight-booking portal (company) need not maintain the testing infrastructure continuously. This results in cost savings for the company as it is using pay-per-use model. The service provider utilizes the same automated test framework for any new carrier.

Hence, TaaS provides advantage to both service provider and the consumer if chosen wisely. Taking one more step, we can add this TaaS solution on cloud. We can use cloud storage for storing the data from the carrier before moving the same to the database.

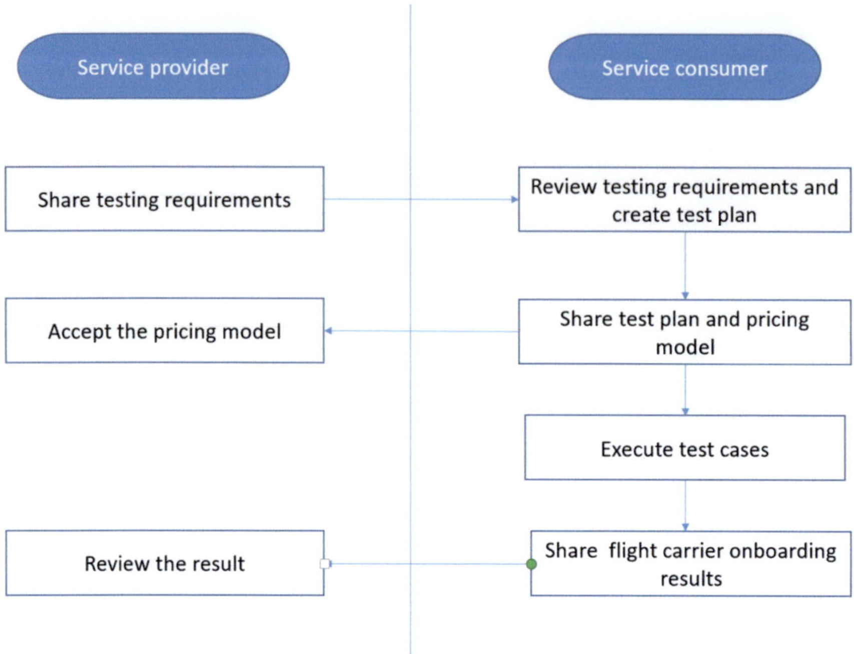

Fig. 23 Interaction between a service consumer (portal) and a provider

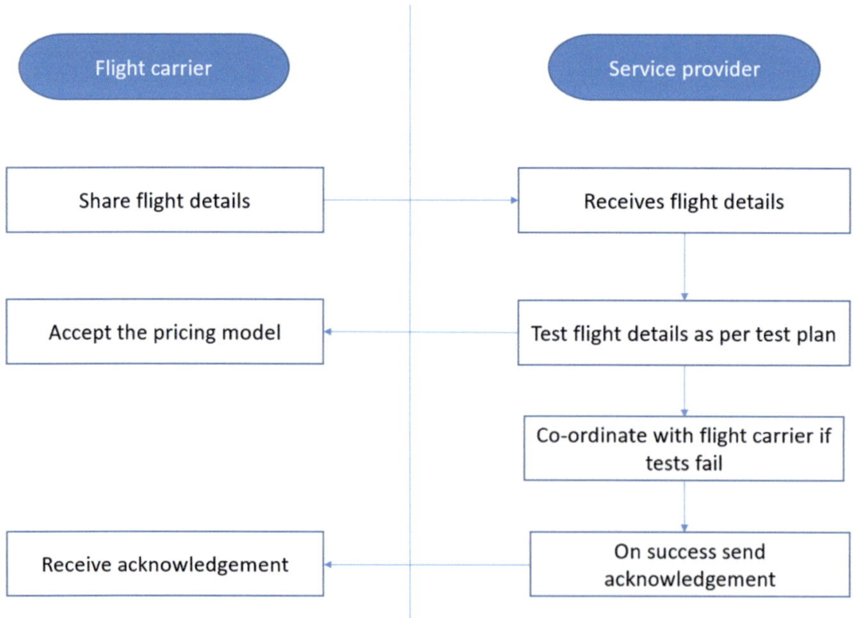

Fig. 24 Interaction between the flight carrier and the service provider

4.2 Implementation

In this section, we present an outline implementing the industry example explained in Sect. 4.1.

As soon as the flight-booking portal places request for testing, the provider collects the test requirements. Figure 25 shows a few test requirements shared by the portal.

The service provider gets the requirements from the service consumer through interface. The interface executes ProcessTestRequirement method to input the requirements into the framework. After requirements input, the interface executes GenerateTestScenario method to generate test scenarios. (Refer Sect. 3.1 for details).

Figure 26 shows test scenarios.

After this, the interface invokes GetScenarioDependent method that generates dependents shown in Fig. 27.

Requirement 1:
Verify the format of data shared by flight-carrier is in the correct format and has all the required information.
Requirement 2:
Store all the information in the flight-booking portal's database.
Requirement 3:
All the data(like flight reschedule and seats booked), should be updated live.
Requirement4:
Portal should work fine on all the versions of popular browsers like IE, safari, firefox and chrome.

Fig. 25 Test requirements

Test scenario 1:
Verify that the data shared by carrier is complete. It contains source, destination, dates, time, fare and seat count.
Test scenario 2:
The user is able to search new carrier's flight details after new carrier is added.
Test scenario 3:
The user is able to successfully book the ticket and the money transition is secured

Fig. 26 Test scenarios

Fig. 27 Scenario dependents

Scenario Dependents :
Operation Systems : Windows 7 and above. iOS
Browsers : IE, firefox, chrome, safari and opera

Test scenario 1: Verify that the data shared by carrier is complete. It contains source, destination, dates, time, fare and seat count.
Test case 1: Verify the input file has source.
Test case 2: Verify the input file has destination.
Test case 3: Verify the input file has date and time of journey.
Test case 4: Verify the input file has seat count.
Test case 5: Verify the input file has fare.
Test case 6: Verify the input file has valid flight id.

Test scenario 2: The user is able to search new carrier's flight details after new carrier is added.
No test cases found in the repository

Test scenario 3: The user is able to successfully book the ticket and the money transition is secured.
Test case 1:Required flight is displayed correctly.
Test case 2:Seats and fares details are correct.
Test case 3:User is able to modify the seats.
Test case 4: User can apply discount coupons.
Test case 5: User is able to do successful payments

Fig. 28 Test cases fetched from test case repository

In parallel, the interface executes FetchTestCase method that fetches all the applicable test cases for these scenarios from the existing repository. Figure 28 shows the output of FetchTestCase method.

Next is ReviewAndUpdate method, which is a manual step. In this step, the tester reviews the test cases and adds the test cases for test scenario 2 for which no test cases are present in the repository. Figure 29 shows the test cases added by the tester.

Finally, the test plan is generated using GenerateTestPlan method. Figure 30 shows a test plan for different test scenarios to be executed in different platforms.

The service provider shares the test plan with the consumer for review. The consumer seeks any clarification as needed and finalizes the test plan. Now, the interface calls Pricing calculator module to estimate test cost and propose to a service consumer. Test service cost depends on testing units each test case consists of. A discussion on test service cost is made in the next section.

As explained in Sect. 3.2, the service provider performs testing after consumer agrees on cost of the service. GetAllTestCases() (Fig. 19) fetches test case code

Test scenario 2: The user is able to search new carrier's flight details after new carrier is added.
Test case 1: Check if the new flight is available for any date.
Test case 2: Other flight are also shown as available.
Test case 3: Check that the flight is not shown as available for the date when all seats are booked.

Fig. 29 Test cases added by the tester

Devices	Browser	Test Cases to be executed	Time taken	Remarks
Windows7	IE, chrome	TestScenario1& TestScenario3	8 hours	
Windows8	IE, safari	TestScrenario2& TestScenario3	6 hours	
Windows8.1	IE, firefox	TestScenario1 & TestScenario3	8 hours	
Windows10	All	All	12 hours	
Mach OSX 10.8	Safari, chrome	TestScenario1& TestScenario3	8 hours	
Mach OSX 10.9	Safari, firefox	TestScenario1& TestScenario2	6 hours	
Mach OSX 10.10	Safari, IE	TestScenario1& TestScenario3	8 hours	
Mach OSX 10.11	All	All	12 hours	

Fig. 30 Test plan

available in the Test case repository, and for other test cases, tester may write test code. TaaSExecutor (Fig. 18) executes all the test cases with respect to a test plan, and corresponding test result report is generated.

In this section, we have extended TaaS framework suggested in Sect. 3 to support an industry example. We hope this brings clarity regarding the framework implementation. We can extend this framework for supporting multiple service providers.

5 Pricing Test Service

A test service provider needs to price its service in a logical manner not only for its cost benefit but also to make a service consumer agreed upon. There are many cost models for assessing software development cost. All these models primarily consider human efforts, problem complexity, and related issues for pricing software. Similar, approach for costing a test service can be considered here.

Methods of costing can be either linear or nonlinear. Linear model calculates cost based on the estimated effort and rate of cost per unit of effort, whereas in the nonlinear pricing model the cost depends on various factors and varies during the product life cycle. The nonlinear model provides flexibility to consumers. A service provider and a consumer explore various options for a task in hand while costing. Costing for test services can adopt any of these two based on quantum of a service. A relatively small in size software testing service requiring not much time can adopt linear model whereas testing of larger size and spanning over long time for various test conditions can adopt nonlinear model for costing. So, for test service costing size of software and its complexity and varied test conditions are the points of considerations.

The steps involved in costing starts with sharing of test requirements to test service provider by a service consumer. For example for updating flight-booking portal, portal agency needs to specify not only the desired updates but also the devices, operating systems, and browsers on which updated system is required to perform. Further security, language support, and other domain-specific issues are to be specified. The clarity in specifying test requirements is of importance here because testing service provider is of separate agency and this separation could be a disadvantage for provisioning a service.

On knowing test requirements, a service provider is required to assess resources it requires to deliver a service in stipulated time. Resource includes both human expertise and tools support. In case of non-availability of required resource, service provider may outsource those for the service in hand. This has bearing on costing because of involvement of third-party service. Thus, TaaS costing can be complicated as it looks here. Unitization of test requirements and finding the cost per unit based on resource requirements are the basic issues that need to be considered while costing a test service. A formal cost model for test service is a matter of research.

6 TaaS on Cloud

In previous sections, we have established that TaaS is a viable testing model to achieve reduced costs and improved services. NASSCOM report [11] expects the worldwide software testing outsourcing market to grow from $30 Billion in 2010 to $50 Billion in 2020. Keeping in view of the emerging business opportunity, test service needs to be engineered in a way so that providers can make good ROIs. This requires optimal usage of resources to reduce service cost. Cloud technology now provides a means for resource sharing. In this section, we explore the usages of cloud technology for TaaS.

6.1 Benefits

While TaaS model allows the usage of several hardware and software platforms for a testing service, a provider may not have such resource on its own. Further, as technology often changes, there could be several versions of these resources. A test service provider acquiring all these versions is also an expensive proposition. In order to meet this challenge, solution can be found with cloud technology on which a service provider can find required resources.

A cloud model provides a simplified [12] solution to manage resource utilization costs since the servers, hardware, and their applications are on cloud. Depending on the enterprise requirements, the amount of infrastructure utilized can be increased or decreased to manage the existing load through the benefits of a pay-per-use model. Consider an application having both web and mobile interface. We need to test this on various laptops, desktops, tablets, and mobiles. Hardware purchase and recurring maintenance costs are significant. In such scenarios, it is advisable to move to cloud and use virtual machines instead of hardware for better ROI.

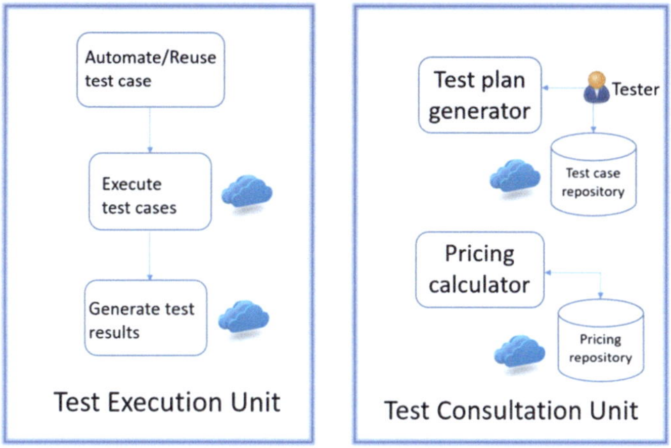

Fig. 31 Modules of TaaS framework on cloud

6.2 *Implementation on Cloud*

We discussed the benefits of moving to cloud. In this section, let us consider moving the TaaS framework proposed in Sect. 3 to cloud. We can move various modules of the proposed framework to cloud. Figure 31 shows an overview of the proposed modules on cloud.

Test case and pricing repositories are two databases that store test cases and test costs, respectively. We can move these repositories to cloud as cloud takes auto backups and provides auto recovery for any hardware or software failure. In addition, as database size grows, scale-up repository facilities can be provided over cloud. Similarly, a service provider can put test service results on cloud so that a service consumer can avail it.

Execute test cases: A test service needing several platforms can avail those on cloud. Further simulators and tools available on cloud can be used in delivering a test service. Such services made available on cloud are used on pay per use [13] basis. In this case, a test service provider does not need to invest for acquiring the resources instead it pays for its use only.

7 Conclusion

"As a service" solutions are gaining momentum. We witness TaaS acceptance in software industry. For smooth transition from traditional testing model to TaaS model, test service automation tool can greatly help. A TaaS automation tool is generally a collaborative web-based portal that offers test planning, test construction, and test case management functions throughout software testing. A customer wishing to avail such a service first has to undertake a well-defined SLA. Then, the required testing service is provided as per the agreed SLA. IBM, Wipro, and HP are a few leading test service providers [14] that offer tools enabling software developers in availing their test services.

On recognizing the recent TaaS developments, the chapter identifies the issues involved and presents a mechanism for implementing a generic TaaS framework. We have explained behavior of the proposed generic framework with the help of an example drawn from real-life application. Industry and academia can further research and improvise the proposed framework. In this chapter, we have presented TaaS architecture, TaaS enablers, factors involved in moving from traditional testing to TaaS, and its benefits and limitations. By moving traditional testing to TaaS, it is possible to optimize testing costs by opting for outsource service and not making a permanent investment for it. The approach has inherent potential for quality because of well-orchestrated collaboration of developers and testers. We further have explored how TaaS on cloud presents cost and efforts optimization opportunity. TaaS is nascent but promising. As it happens with any emerging solutions, standard universally agreed options are missing. Big technology players can research and innovate more in TaaS to realize its true potential and standardize available solutions and frameworks.

References

1. http://www.academia.edu/8477067/Testing_as_a_Service_on_Cloud_A_Review
2. https://en.wikipedia.org/wiki/Software
3. http://en.wikipedia.org/wiki/Chinook_helicopter
4. https://www.ima.umn.edu/~arnold/disasters/ariane5rep.html
5. www.nist.gov/director/planning/upload/report02-3.pdf
6. www.cs.toronto.edu/~jm/340S/02/PDF2/Lifecycles.pdf
7. http://www.softwaretestinghelp.com/types-of-software-testing/
8. http://blog.qatestlab.com/2011/02/26/testing-as-a-service-outsourcing-your-specialized-software-testing/
9. http://www.nasscom.in/software-testing-emerging-opportunities
10. https://en.wikipedia.org/wiki/Service_(economics)
11. www.cognizant.com/InsightsWhitepapers/Taking-Testing-to-the-Cloud.pdf
12. http://finance.flemingeurope.com/webdata/4899/LeanApps_White_Paper_Testing_as_a_Service_ENG.pdf
13. http://www.csc.com/independent_testing_services/offerings/82906/83166-testing_as_a_service_taas?ref=lc
14. http://www.softwaretestinghelp.com/software-testing-service-providers/

MIX
Papier aus verantwortungsvollen Quellen
Paper from responsible sources
FSC® C105338

If you have any concerns about our products,
you can contact us on
ProductSafety@springernature.com

In case Publisher is established outside the EU,
the EU authorized representative is:
Springer Nature Customer Service Center GmbH
Europaplatz 3, 69115 Heidelberg, Germany

Printed by Libri Plureos GmbH
in Hamburg, Germany